STP 1071

Biological Contaminants in Indoor Environments

Philip R. Morey, James C. Feeley, Sr., and James A. Otten, editors

ASTM
1916 Race Street
Philadelphia, PA 19103

Library of Congress Cataloging-in-Publication Data

Biological contaminants in indoor environments / Philip R. Morey, James C. Feeley, Sr., and James A. Otten, editors.
(STP ; 1071)
Contains papers presented at a symposim held in Boulder, July 16-19, 1989, sponsored by ASTM Committee D-22 on Sampling and Analysis of Atmospheres and its Subcommittee D22.05 on Indoor Air, Section .06 on Biological Aerosols.
"ASTM publication code number (PCN) 04-010710-17"--T.p. verso.
Includes bibliographical references and indexes.
ISBN 0-8031-1290-4
1. Air--Microbiology--Congresses. 2. Indoor air pollution--Congresses. 3. Air quality management--Congresses. I. Morey, Philip R. II. Feeley, James Corrigan, 1937- . III. Otten, James A., 1939- . IV. ASTM Committee D-22 on Sampling and Analysis of Atmospheres. V. ASTM Committee D-22 on Sampling and Ayalysis of Atmospheres. Section .06 on Biological Aerosols. VI. Series: ASTM special technical publication ; 1071.
QR101.B56 1990
628.5'36--dc20 90-2111
CIP

NOTE

The Society is not responsible, as a body, for the statements and opinions advanced in this publication.

Peer Review Policy

Each paper published in this volume was evaluated by three peer reviewers. The authors addressed all of the reviewers' comments to the satisfaction of both the technical editor(s) and the ASTM Committee on Publications.

The quality of the papers in this publication reflects not only the obvious efforts of the authors and the technical editor(s), but also the work of these peer reviewers. The ASTM Committee on Publications acknowledges with appreciation their dedication and contribution of time and effort on behalf of ASTM.

Printed in Baltimore, MD
December 1990

Foreword

This publication, *Biological Contaminants in Indoor Environments*, contains papers presented at the symposium of the same name held in Boulder, CO, on July 16–19, 1989. The symposium was sponsored by ASTM Committee D-22 on Sampling and Analysis of Atmospheres and its Subcommittee D22.05 on Indoor Air, Section .06 on Biological Aerosols. Dr. Philip R. Morey of Clayton Environmental Consultants, James A. Otten of Martin Marietta Energy Systems, Inc., and the late James C. Feeley, Sr. of Pathogen Control Associates, Inc., presided as symposium chairman. They also served as editors of this publication.

James C. Feeley, Sr.
1937-1989

Dedication

The proceedings of this symposium are dedicated to the memory of James C. Feeley, Sr. (1937-1989). Jim was involved in the planning and organization of this symposium and is listed as one of the editors for this important work on biological contaminants in indoor environments.

Jim gave abundantly of his time, talent, and knowledge to further the science of microbiology and the emerging field of indoor air quality. Many of us are beneficiaries of his constant efforts to strive for excellence; the science of microbiology is his debtor.

Jim Feeley received his Ph.D. in microbiology from the University of Georgia, Athens in 1978. Jim had already begun his distinguished career with the Centers for Disease Control (CDC) in Atlanta which continued until his retirement in 1986. Jim began his career as a microbiologist with CDC in 1964 and at the time of his retirement he was Chief of the Field Investigations Laboratories, Respiratory and Special Pathogens Branch. During this time, Jim authored or co-authored over 100 publications. His earlier publications included work on Bacillus anthracis. In 1969 Jim was part of a team that performed an epidemiologic study of inhalation authrax. Later publications were centered around Yersinia enterocolitica.

Following the outbreak of Legionnaires' Disease in 1976, Jim played an active role in studying the epidemiology of the disease and the growth characteristics of Legionella pneumophila. Jim received a special citation for his work on Legionella by the Infectious Disease Society of America and in recognition for his work with Legionnaires' Disease, the species Legionella feeleii was named in his honor. In addition to these honors, Jim also received six special commendations from the directors of the CDC.

Following his retirement from CDC, Jim formed his own company, Pathogen Control Associates, Inc., where he remained until his sudden death from a heart attack on September 15, 1989. Jim's company provided a valuable consulting service to other organizations in the identification of microbial contaminants and their subsequent remediation. It is an indication of Jim's foresight that this company continues to provide this service today.

Jim always found time to give sage counsel to professional societies. He was a charter member of the American Conference of Governmental Industrial Hygienists' (ACGIH), Bioaerosols Committee. His thinking on "reservoirs, amplifiers, and disseminators" is evident in the latest "Guidelines for the Assessment of Bioaerosols in the Indoor Environment" published by the ACGIH. Jim's counsel will be missed by ventilation engineering colleagues. He was a charter member of the American Society of Heating, Refrigerating, and Air-conditioning Engineers', Inc. (ASHRAE), Environmental Health Committee. The latest ASHRAE position paper on Legionnaires' Disease largely derives from Jim's thoughts and input.

We honor and cherish the memory of our departed colleague. We will miss his encouragement, counsel, hearty laughter, and good nature. We dedicate ourselves and this volume to his memory and to the attainment of the objectives that Jim held so dear.

James A. Otten
Martin Marietta Energy Systems Inc.
Oak Ridge, Tennessee 37831

Philip R. Morey
Clayton Environmental Consultants
Wayne, Pennsylvania 19087

Contents

Overview

The purpose of this symposium on biological contaminants in indoor environments was to develop and explore sampling and analytical protocols for microbial agents that may be commonly or uncommonly found indoors. Classes of microbial agents considered in this symposium were:

- Viruses
- Bacteria including, gram negative, gram positive, Legionella, and mycobacteria
- Specialized bacteria including Chlamydia and rickettsia
- Fungi including saprophytes and pathogens
- Protozoa
- New microorganisms
- Mycotoxins and endotoxins

This symposium was organized by ASTM Committee D22.05.06 (Indoor Air, Biological Aerosols) because of increasing concern over possible involvement of microorganisms in building related illnesses and sick building syndrome. The extent to which microorganisms are involved in building sicknesses is still unknown. Sampling and analytical protocols for most types of microbial agents are poorly developed and in those few instances where protocols are available, interpretation of analytical results is inconsistent.

The volume begins with a review by Morey and Feeley on the importance of microbial agents in indoor air quality issues. Twelve chapters follow, each on a specific class of microorganism or microbial agent. Emphasis in these chapters is placed on sampling and analytical protocols, case studies, and data interpretation. An important aspect of each of these 12 chapters is the discussion and closure session where authors responded to the specific questions of the attendees of the conference held in Boulder, Colorado on July 16-19, 1989. The last three chapters provide viewpoints on future developments in the field of indoor bioaerosols research and the likely future involvement of bioaerosols in building performance.

Most of the technical data presented at this conference is primarily of use to microbiologists and professionals who investigate environmental problems in indoor environments. Each of the 12 chapters on specific types of microbial agents provides information on reservoirs where the agent may accumulate, conditions under which the agent may grow or amplify, and examples where dissemination to the breathing zone occurs. Some emphasis is also given to the possible adverse health effects caused by each agent as well as to interpretation of analytical results. Many of the questions on microbial contaminants that are often asked by occupants and owners of buildings are answered in the discussion and closure section at the end of each chapter.

Hierholzer discussed the biological properties and the diseases caused by viruses. The diseases involve mainly the respiratory, conjunctivial, and gastrointestinal tracts of affected individuals.

The physical makeup of viruses determines their ability to survive in the environment with the majority being highly susceptible to desiccation. Viruses can be collected from the air by impaction onto agar surfaces or into liquid media with subsequent growth in cell culture. In addition to cell culture, some viruses can be detected directly by fluorescent antibody tests, enzyme assays, and DNA probes.

LaForce, Hood, Falkinham, and Dennis described gram positive, gram negative, acid fast, and legionella bacteria. Several gram positive bacteria can be recovered from room air, but in cases involving micrococci and staphylococci, the relevance of their role as pathogens has not been established. The pathogenicity of Bacillus anthracis, the causative agent of anthrax, is well known and is used as a case study in describing the transmission of gram positive organisms by the aerosol route. Gram positive bacteria can be collected by impaction on agar and by impingement in liquid media with subsequent incubation at 37C or at room temperature for 48 hours. Filter cassettes can also be used to collect gram positive bacteria, but care must be taken to prevent desiccation which destroys viability in most cases.

Gram negative bacteria have been associated with respiratory diseases and other syndromes where exposure involved bioaerosol emissions from heating, ventilation, and air-conditioning and humidifying systems. An outbreak of hypersensitivity pneumonitis, possibly caused by Cytophaga, is presented as a case study. All gram negative bacteria possess an endotoxin (see chapter by Olenchock) in their cell wall and endotoxin is an important factor in respiratory diseases. Gram negative bacteria can be collected by impaction on agar, in liquid media, and also by filter cassettes. As with gram positive bacteria, gram negative bacteria can be destroyed by desiccation on filters. The organisms are grown on agar media at room temperature, 35C, or 45-50C for evaluation. There is still a lack of information regarding the exact role of these bacteria in respiratory diseases.

Mycobacteria are known to cause disease by the respiratory route; especially well studied is tuberculosis caused by Mycobacterium tuberculosis. Other mycobacteria occurring in natural aerosols do cause human infection, but their epidemiology and route of human infection is less well understood. Mycobacteria can be recovered using impaction on specialized media with subsequent growth at 37C for up to eight weeks.

Legionella pneumophila, causative agent of Legionnaires' disease, is a common aquatic organism which is able to colonize manmade water systems. It is classed as an opportunistic pathogen for man and causes a pneumonia which can be fatal. Pontiac fever, a non-pneumonic disease, has also been linked to Legionellae but is more influenza like. Legionellae can be isolated from a wide range of habitats, especially from warmer waters (35-60C). The Legionellae are generally collected in bulk water samples, concentrated, and their presence detected by fluorescent antibody tests. Legionellae can be identified by growth on special media by incubation at 35C for several days. Unlike most other airborne organisms, Legionellae are difficult to collect in aerosols.

Cole and Regnery discussed two specialized bacterial groups, the _Chlamydia_ and rickettsia. The _Chlamydia_ consist of three species of organisms causing diseases including pneumonia, conjunctivitis, and diseases of the reproductive and urinary tracts. The organisms replicate only in the cytoplasm of eucaryotic cells and are harbored by man and other animals, especially birds. All three species can transmit disease by the aerosol route. Recommended sampling methods include the liquid impinger and membrane filtration. Chlamydia are analyzed by utilizing a direct fluorescent antibody test or grown in cell culture. Rickettsia are thought to be transmitted between vertebrate hosts exclusively by arthropod vectors. However, the agent responsible for Q fever, _Coxiella burnetii_, can occur in aerosols and the organism is extremely infectious by inhalation. Most disease in humans is inapparent or self-limiting. However, life-threatening endocarditis can occur. Evidence suggests that _Coxiella burnetii_ may be a common environmental contaminant when appropriate animal reservoirs are present. Cattle, sheep, goats, and several wild animal species harbour the agent. _Coxiella burnetii_ can only replicate within living cells. The agent is sensitive to tetracycline and vaccines are available. The agent can be collected easily using liquid impingers or cotton filters. _Coxiella burnetii_ is metabolically inactive at any pH other than the acid environment associated with the phagolysosome (pH 4.5).

Burge described the saprophytic and pathogenic fungi. Fungi are heterogenous organisms grouped by structure, biochemistry and physiology. Most fungi are able to use non-living organic material as a substrate, but a few are pathogenic and will invade human tissues. Fungi can cause human hypersensitivity, infections, and toxic diseases. Infectious diseases are generally of four types: cutaneous, subcutaneous, systemic, and those that cause opportunistic infections when host defenses are impaired. Cutaneous and subcutaneous mycoses are not considered airborne diseases. Fungi produce a variety of antigenic materials that are spread by the aerosol route. Sensitization to the antigens can occur and result in allergenic diseases such as asthma and hypersensitivity pneumonitis. Fungi need not be viable to be antigenic. In addition, fungi can produce mycotoxins (see chapter by Jarvis) and irritants (for example, volatile organic compounds). Assessment of the indoor environment for fungi should take into account the following: a reservoir that contains living fungi, an amplifier, and a means of dissemination. During an evaluation of a building for indoor environmental problems, bulk samples can be taken to confirm the presence of fungal reservoirs and amplifiers. Air sampling, in general, should only be done to confirm that amplifiers are producing aerosols. Fungi can be collected by impaction onto agar surfaces, into liquids (use a wetting agent because of hydrophobicity), or the spores can be collected on a sticky surface for enumeration. Sample analysis depends on the mode of collection with culture plate samples being incubated for at least 5 days at room temperature for saprophytes and 37C for human pathogens. Spores collected on sticky surfaces are enumerated by examining them microscopically. Fungi can also be collected on filters and the material enumerated by culture on suitable media or by counting the spores directly. Antigens are assayed using immunological methods. At the present time it is impossible to assess numeric risk for airborne fungi.

Tyndall discussed the health concerns associated with protozoa in indoor environments. Free living amoeba are the likely protozoa that could be implicated in health concerns in the indoor environment. Amoeba can cause allergic reactions, eye infections, and encephalitis. Amoeba can support the multiplication of Legionella. Because of their large size, free living amoeba are not easily aerosolized so they are usually isolated in bulk samples. If air sampling is done, impaction onto agar or collection into liquid is used. Detection methods include plating on agar with E. coli or inoculation into animals. These methods are laborious and time consuming, but new techniques such as monoclonal antibodies and gene probes exist for rapid detection.

Plouffe presented a historical review of organisms causing respiratory infections, especially pneumonia. This discussion centered around the fact that respiratory infection in each recent decade was associated with a particular class of organism and the realization that new organisms would arise as causative agents of pulmonary infections. Future research should concentrate on new culture media, immunocompromized patients, epidemics, and environmental isolates.

Olenchock discussed endotoxins which are an intergral component of the outer membrane of gram negative bacteria. Endotoxins are potent biological agents that can be associated with acute lung reactivity in exposed individuals. They are found in the soil, water, and in other living organisms around the world. Agricultural environments are a prime source, but they can also be identified in office buildings with humidification systems that contain water sumps. Endotoxins can be sampled from bulk materials, water, and airborne dusts. Filters are used commonly to collect dusts and the sample is analyzed using the Limulus amebocyte lysate gelation test. There are newer test methods under development.

Jarvis described mycotoxins and their association with indoor air quality. Mycotoxins are secondary metabolites produced by fungi. Cases of intoxication through ingestion of food have been reported; however, little is known about the threat of airborne toxigenic fungal spores. Mycotoxins may play a role in the symptoms experienced by occupants of buildings that are heavily contaminated by certain species of fungi. Fungal spores need not be viable to elicit a toxic effect so sampling techniques should employ samplers that collect viable spores (impaction onto agar with subsequent incubation) and non-viable spores (filters, spore traps). Species identification is important because of the large variation among taxa of the same genus in the production of mycotoxins. Air samples must be collected on several different occasions in areas where complaints are the highest and also in areas where there are no complaints. Analysis can take the form of injection of material into susceptible animals or if the microbial and toxicological analysis point to a particular fungus whose toxins are known and available, as standards, the samples can be analyzed by a variety of chromatographic, spectometric, and immunoassay techniques.

In the last 3 chapters in this volume Burge, Feeley, Morey, and Levin describe future bioaerosol problems in buildings and the intervention actions and studies that will likely result from attempts to reduce indoor microbiological pollution. Feeley and Morey describe how bioaerosol problems indoors will likely increase in importance in the future. Poor preventive maintenance of building systems is often cited as the major reason for the increase of microbial contamination indoors. Studies will be carried out in the future to determine if intervention techniques such as more effective filtration, increased outdoor air ventilation, and better accessibility of ventilation system components for purposes of maintenance result in reduced concentrations of indoor microbial contaminants. The incorporation of documented microbial intervention techniques into building codes will be the likely result of these research efforts. As discussed by Levin, architects, building owners, and tenants will consider the cost of microbial intervention procedures versus the possible adverse health effects and productivity losses associated with exposure to microbial agents. Burge describes how a significant amount of future study will be directed toward establishment of exposure/dose-response relationships for common diseases caused by bioaerosols indoors. The possible development of bioaerosol exposure guidelines will result from these efforts.

Funds to support the travel of speakers to the Boulder Conference on Biological Contaminants in Indoor Environments were provided by the United States Environmental Protection Agency. The co-editors and ASTM are grateful for the Environmental Protection Agency's support of this scientific endeavor.

Philip R. Morey
Clayton Environmental Consultants
Wayne, Pennsylvania 19087

James A. Otten
Martin Marietta Energy Systems, Inc.
Oak Ridge, Tennessee 37831

Philip R. Morey and James C. Feeley, Sr.

THE LANDLORD, TENANT, AND INVESTIGATOR: THEIR NEEDS, CONCERNS AND VIEWPOINTS

REFERENCE: Morey, P. R. and Feeley, J. C., Sr., "The Landlord, Tenant, and Investigator: Their Needs, Concerns and Viewpoints," Biological Contaminants In Indoor Environments, ASTM STP 1071, Philip R. Morey, James C. Feeley, Sr., James A. Otten, Eds., American Society for Testing and Materials, Philadelphia, 1990.

ABSTRACT: Microorganisms in indoor environments may cause adverse health effects that are either infective or allergic in nature. The current state of knowledge of the microbiologist or investigator dealing with air quality problems is reviewed. Examples taken from the literature and recent case studies suggest that a wide variety of microbial agents can adversely affect air quality in interior environments. Sampling and analytical protocols for most kinds of microbial agents are poorly developed. The interpretation of microbial sampling data is far from certain especially from the viewpoint of building owners and tenants. Examples given in this review introduce the theme of the Boulder Conference on Biological Contaminants in Indoor Environments, namely exploring and developing of sampling and analytical protocols for a variety of agents including viruses, bacteria, fungi, protozoa, and microbial toxins.

KEYWORDS: microbial sampling, indoor air quality, data interpretation, viruses, fungi, bacteria

In recent years, microbiological problems in indoor environments have received considerable attention. The Legionnaires' Disease outbreak in Philadelphia in 1976 is probably the most publicized case of illness caused by an infective microbial agent in the indoor environment. Other infective agents, such as viruses, that cause acute respiratory illness are probably more prevalent in the indoor environment especially if occupant density is high and maximum recirculation of ventilation air occurs.

Dr. Morey is Director of Indoor Air Quality Services, Clayton Environmental Consultants, Inc., 151 S. Warner Road, Suite 235, Wayne, PA 19087. Dr. Feeley (deceased September 15, 1989) was President of Pathogen Control Associates, 2204 Hanfred Lane, Suite 100, Tucker, GA 30084.

During the past three or four decades, significant changes have occurred in the construction and operation of commercial buildings. The building envelope has become tighter, and less outdoor air is used in heating, ventilation, and air-conditioning (HVAC) systems. Microbial agents that may be shed by occupants are not so readily diluted by mixing with outdoor air. Energy conservation programs may result in the buildup of moisture in indoor environments that facilitate the growth of certain kinds of microorganisms. As a consequence of neglected maintenance programs, excessive dirt and debris (potential nutrients for microorganisms) accumulate in niches in the HVAC system that provides ventilation air to occupants.

In the past decade, a number of occupant maladies referred to by various terms such as building related illness, building sickness, and sick building syndrome have been generally recognized. Agents such as volatile organic compounds, environmental tobacco smoke, and bioeffluents are often cited as causes of these maladies. Microorganisms or their components may be involved in the etiology of building related illnesses that are either infective or allergic in nature. Microbial volatiles, irritants, and toxins are potentially the causes of some cases of building sickness.

At present, the extent to which microorganisms or their components are involved in the etiology of building sickness is not known. Sampling and analytical protocols for most kinds of microbial agents are poorly developed, and in the few instances where protocols are available (for example, saprophytic fungi, Legionella species) data interpretation is inconsistent.

An objective of this conference is to develop and explore microbial sampling and analytical protocols for 12 kinds of microbial agents. Interpreting sampling and analytical data in terms of possible health effects is another conference objective.

Although most of the technical data presented at this conference will be of use primarily to microbiologists and professionals who investigate environmental problems in interior environments, some emphasis will be given to viewpoints, concerns, and needs of the building owner and tenant who must on occasion deal with a microbial environmental problem.

This chapter describes the types of microbiological agents that can be found in the air especially in indoor environments and considers the increasing importance of certain related technical issues. It reviews the general methods of collecting airborne microorganisms. This chapter also presents case studies, a brief interpretation of data, and concludes with a consideration of the needs, concerns, and viewpoints of landlords and tenants who are sometimes confronted with an indoor microbial pollution problem.

Based on the technical data presented by other authors at the conference and the future changes likely in the way that buildings are constructed and operated, we have also speculated on future developments that may occur in the indoor bioaerosol field [1].

MICROBIOLOGICAL AGENTS

A wide variety of types of microorganisms can be found in the indoor environment such as fungi, bacteria, viruses, and protozoa. In the past, most attention has been given to saprophytic fungi and bacteria, probably because many of these microorganisms are easily sampled and enumerated. Agents such as viruses, rickettsia, chlamydia, protozoa, and many of the pathogenic fungi and bacteria are more difficult to culture, and therefore sampling methodology is not yet readily available.

WHERE MICROORGANISMS ARE FOUND

Microorganisms are a normal and essential component of the earth's terrestrial and aquatic ecosystems. Bacteria and fungi break down the complex molecules found in dead organic materials from animals and plants and recycle minerals and carbon to simple substances such as carbon dioxide and nitrates. Microorganisms such as saprophytic (derive nutrients from non-living materials in the environment) bacteria and fungi are commonly found in the soil and the atmosphere.

Saprophytic fungi such as Cladosporium are almost always found in the outdoor air. The concentrations of Cladosporium spores as well as other microbial agents in the outdoor air vary enormously and are influenced by factors that include the amount of vegetable substrate available for spore production, air turbulence, the occurrence of rain, and the time of the day [2]. Cladosporium spores are also found normally in the indoor environment, depending on the amount of outdoor air that either infiltrates into interior spaces or is brought into the HVAC system.

Bacteria that are saprophytic and shed by humans are normally present in increased numbers in indoor environments. Thus, airborne Streptococcus species occur indoors at concentrations one to two orders of magnitude higher than those found in outdoor air [3].

Although microorganisms are normally present in indoor environments, excessive moisture in some interior niches is associated with an increase of microorganisms that makes the interior environment microbiologically atypical.

In residences where poorly maintained humidifiers are a component of forced air heating systems, concentrations of some fungi and actinomycetes (actinomycetes are a type of gram positive bacteria characterized by branching filaments) may become elevated compared to those in the outdoor air. In large commercial buildings, reservoir and amplification sites for fungi, bacteria (including actinomycetes), protozoa, and even nematodes have been documented in water spray sumps and humidifiers of HVAC systems [4,5]. Outbreaks of illness among occupants have been associated with microbial amplifiers in these buildings.

IMPORTANCE OF MICROORGANISMS IN AIR QUALITY

Several recent studies suggest that microbiological air contaminants may be more important than previously recognized. A study of building sickness in 42 buildings in Great Britain showed that illness symptoms increased substantially when the ventilation air was "air-conditioned" (chilled) or humidified [6]. Therefore, the presence of wet niches in air-conditioning systems and the addition of moisture to the air by humidification systems appear to be associated with increased incidence of building sickness. The extent to which airborne microbials are correlated with building sickness was not resolved in these British studies.

An epidemiological study at four Army training centers over a 47-month period showed that the rate of acute respiratory disease was 50% higher among recruits in mechanically ventilated versus naturally ventilated (open windows) barracks [7]. HVAC systems were therefore implicated as a risk factor for the airborne transmission of infection among susceptible occupants in nonindustrial indoor environments.

In a recent study [8] a panel of judges quantified the perception of air quality in 20 randomly chosen buildings in Copenhagen, Denmark. None of the buildings had previously been identified as having air quality problems. Each building was visited by a panel of judges when it was unoccupied and unventilated, occupied and unventilated, and occupied and ventilated. Each judge evaluated both the odor intensity in the building and the acceptability of the air quality. Using this approach, almost half of the pollutant sources in the buildings were determined to be in the HVAC system.

Although specific pollutants from the HVAC system associated with increased odor intensity and more unacceptable air quality were not identified in the Danish study, microorganisms or their products might be involved. Table 1 lists potential sources of microbiological contaminants in HVAC systems that may be associated with the perception of poor air quality.

HEALTH EFFECTS

Microorganisms in indoor environments may cause adverse health effects that are either infective or allergic in nature. The same microorganism (for example, _Aspergillus fumigatus_) may cause both infection or allergy, depending primarily on the susceptibility of the host.

In the following two sections, some representative infectious and allergic microbial agents that may be found indoors are described. Sampling and analytical protocols and deficiencies in current approaches are discussed.

Infective Agents

Legionella pneumophila is a gram negative bacterium that the public most closely associates with infectious disease in buildings.

Legionnaire's disease is an infection that can result in severe pneumonia and requires the patient to be hospitalized and given appropriate antibiotic therapy [9].

Cooling towers, evaporative condensers, potable water, and hot water service systems can be sources or amplification sites for Legionella. However, for Legionella to be infective it must be disseminated from the amplification site to the breathing zone of the susceptible host. Cooling tower drift containing Legionella can become entrained into HVAC system makeup air intake and disseminate into the indoor air. Legionella can also enter indoor air by aerosolization of contaminated water by shower heads or other water spray devices [9].

In spite of the great number of outbreaks of Legionnaires' disease that have occurred since the epidemic in Philadelphia in 1976 [10], experts disagree on concentrations of Legionella species (and serotypes) in cooling towers and other water reservoirs that are considered acceptable. Sensitivity and specificity of test methods for Legionella are not uniformly accepted.

Some building designers and architects continue to locate cooling towers close to outdoor air intakes and other building inlets, even though overwhelming evidence suggests that potential Legionella reservoirs should be located as far as possible downwind from these inlets. Providing an adequate spatial separation of outdoor air inlets from cooling towers reduces but does not eliminate the possibility that aerosol containing Legionella will be entrained in the HVAC system.

Mycobacterium avium is an acid-fast bacterial pathogen commonly recovered from patients with acquired immunodeficiency syndrome. This bacterium has been isolated from both chlorinated and nonchlorinated drinking water supplies. A recent study showed that the relatively low numbers of Mycobacterium avium present in drinking water can become selectively amplified in hot water service locations, including faucets and shower heads [11]. Droplet nuclei from these hot water service sources, especially showers, likely increase the risk that susceptible individuals may have to this bacterium.

Although air sampling for Mycobacterium avium has not been performed in relation to dissemination from hot water amplification sources, the use of heterotrophic agar [11] in a high flowrate impaction sampler might be an appropriate approach. Analytical procedures are available for the isolation of Mycobacterium avium from water samples.

Outbreaks of infective illness in the indoor environment may be caused by other types of microorganisms including viruses. An outbreak of influenza involving 72% of the passengers on a jet airliner was associated with a viral aerosol [12]. Infection occurred during the late winter when passengers together with an acutely ill person (fever, chills, cough) remained onboard a disabled aircraft which was parked for several hours while its engines were being repaired. The plane's ventilation system which normally delivers outdoor air to passengers had been turned off during repairs.

During this period, droplet nuclei from the acutely ill person were likely inhaled by other passengers in the confined, stagnant, and dry interior space. Air sampling was not performed, but throat swabs from several ill passengers analyzed by inoculating embryonated chicken eggs and primary rhesus monkey kidney cells suggested infection with influenza A virus. If air sampling for influenza A virus had been performed, samples containing the virus probably would have been collected throughout the passenger cabin because droplet nuclei from the acutely ill person would likely have been distributed throughout the confined interior space.

Q fever is a febrile illness caused by inhalation and infection with a rickettsial organism, _Coxiella burnetii_. The usual reservoirs for this organism are domestic cattle, sheep, and goats. A recent outbreak of this illness involving 33 cases in the small town of Baddeck, Nova Scotia, was associated with exposure to a parturient cat [13]. The cat lived in one of four buildings where 14 of the cases were likely exposed to the disease agent.

Coxiella burnetii is a desiccation-resistant bacterium that is shed from animal reservoirs, such as urine, feces, and especially birth products. Presumably a filter cassette sampler would have been the appropriate sampling probe for this microbial agent had the investigators [13] attempted to demonstrate the exact pathway(s) of airborne transmission.

Histoplasmosis, an illness caused by the fungus _Histoplasma capsulatum_, is a chronic lung disease that an individual may develop when exposed to conidia (asexual spores) of this organism. Exposure to conidia in indoor environments can occur when, for example, bat droppings from attics of abandoned buildings are removed by workers who are not wearing respirators [14]. _Histoplasma capsulatum_ is usually detected by analysis of environmental reservoirs, such as contaminated soil. There is little or no data available on sampling to demonstrate the airborne transmission of this pathogenic fungus.

Allergic Agents

Microorganisms may cause adverse health effects in indoor environments by affecting the immune system. Allergic respiratory diseases are caused by a hypersensitivity response to inhaled particles containing microorganisms and their components, such as spores, toxic metabolites, enzymes, and cell wall fragments. In Europe and North America, there have been numerous reports of building related illness (often called humidifier fever or humidifier disease in Europe; hypersensitivity pneumonitis in North America) in which affected individuals manifest acute symptoms, such as malaise, fever, shortness of breath, cough, and muscleaches [4,5,15]. These illnesses usually occur as responses to microbiological contaminants aerosolized from HVAC system components, such as humidifiers and water spray units, or from other components of buildings that may have been damaged by recurrent floods or moisture [16,17]. Affected individuals in office environments usually experience relief when they leave the building for an extended period of time.

Common environmental microorganisms, such as Cladosporium and Penicillium, may be etiologic agents of hypersensitivity lung illness. For example, exposure to Cladosporium growing on a wooden ceiling in a residential hot tub room caused hypersensitivity pneumonitis in one individual [18]. Presumably, the affected occupant inhaled massive doses of this otherwise common environmental microorganism.

Many additional microbial agents including thermophilic actinomycetes [19], Cytophaga (a gram negative bacterium) species endotoxin [20], and nonpathogenic free living amoebae [5] may cause hypersensitivity lung illness. Air sampling methods for some of these etiologic agents such as amoebal cysts and antigen have not been described in the literature.

CASE STUDIES

Five examples follow (three in Table 2) where sampling for microorganisms was performed in office environments. In some examples, microbiological analytical results are easily interpreted. In other cases, the interpretation of microbiological sampling data is problematical.

Building A - Fungi and Actinomycetes From Wet Sound Liner

Air sampling for fungi and thermophilic (grow at 50 to 55 °C) actinomycetes was performed in office zones of a building where a significant number of occupants reported sick building syndrome symptoms and odor annoyance [21]. Cladosporium dominated the fungal flora in the air outdoors (Table 2). In offices, the total concentrations of fungi were about 10 times the outdoor concentration, and Penicillium dominated the indoor flora. Thermophilic actinomycetes were found in the indoor air in a few offices at concentrations up to 100 times those outdoors (Table 2). Microbial populations were clearly being amplified in Building A. Further evaluation showed that Penicillium and thermophilic actinomycetes were amplifying on the wet, porous manmade insulation (sound liner) in several air handling units that provided ventilation air to the occupied space.

Results of the samples collected in Building A successfully demonstrated the presence of elevated bioaerosol populations because the amplifier (source) was moist and the HVAC system was transporting spores to the occupied spaces [22].

Building B - Fungi From Dry Sound Liner

Air sampling for fungi was performed in an office building where the sound liner in air handling units was encrusted with a "moldy" layer of dry debris (Table 2). Samples from offices showed that concentrations of fungi were less than 10% of those outdoors. When dry sound liner was disturbed during infrequent maintenance activities, air containing greatly elevated concentrations of spores (up to 200 times those outdoors) was present in offices.

In Building B, sampling at most times in offices, in absence of a thorough inspection of the HVAC system, would not have detected the presence of fungal amplification sites in the HVAC system.

Building C - Microorganisms In Room With Portable Humidifier

Air sampling was performed in an office where floods had previously wetted several ceiling tiles (Table 2). Concentrations of fungi in offices during sampling were less than 10% of those outdoors. When damaged ceiling tiles were disturbed (slight movement), concentrations of fungi in offices increased to about 13 times those outdoors. Sampling performed when ceiling tiles were disturbed effectively demonstrated the presence of a dry microbial reservoir.

During the course of air sampling in Building C, it was determined that one occupant in the office was undergoing antibiotic treatment for a respiratory infection caused by Pseudomonas aeruginosa (a gram negative bacterium). Inspection of this office revealed that a portable humidifier was located about 2 meters from the occupant's breathing zone.

Sampling analysis for airborne bacteria (Table 3) showed that low numbers of Bacillus (a gram positive bacterium) were recovered from office air when the portable humidifier had been turned off for half a day or more. Concentrations of airborne bacteria increased in office air within a few minutes after turning on the humidifier (Table 3). Some Pseudomonas (but not P. aeruginosa) were found in the humidifier aerosol (Table 3; see reference 9). The water samples from the humidifier reservoir, however, showed that Pseudomonas aeruginosa was the predominant bacterium present in this device. The probable source of the Pseudomonas aeruginosa causing the occupant's respiratory infection was the portable humidifier.

Bioaerosol studies in Building C illustrated the following points:

1. The availability of medical data on occupant illness is critical in deciding which microorganism or class of microorganism should be collected.

2. Some microorganisms (Legionella is another example) are not readily collected by air sampling. "Bulk" water samples in Building C proved to be more useful in finding the cause of illness than air samples.

3. Air sampling for saprophytic fungi (Table 2) was of little or no use in finding the source of this occupant's illness.

Building D - Bacteria In a Crowded Office Building

Sampling for airborne gram positive bacteria was performed in a crowded office with many smokers and considerable environmental tobacco smoke [9]. The bioaerosol analytical data shown in Table 4 suggests that Micrococcus found in the office probably originated in the outdoor environment. Staphylococcus aureus and S. epidermidis, both of which are common human shed bacteria, are highly represented among the gram positive bacteria found in office air. Gram positive bacteria found in this crowded office (about 10 to 14 occupants/1000 square feet) are probably only a crude indicator of the density of human occupancy.

No conclusion on risk of infection or other adverse health effects from these bacteria should or can be made because we are not dealing with immunocompromised individuals or critical care facilities (hospital) where these microorganisms are traditionally a cause for concern. Sampling for bacteria in this building should not have been performed because there was no evidence relating bioaerosols to complaints and because, from the onset of the evaluation, it was apparent that smoking in offices was excessive and complaints (headache, odor annoyance) were probably due to environmental tobacco smoke.

Building E - Humidifier Fever Outbreak

More than 25% of office occupants in a zone of a large building exhibited symptoms of humidifier fever. Illness was associated with chronic floods and water disasters [16]. Water damaged furnishings and dust and debris from the HVAC system contained a wide variety of microorganisms including Acanthamoeba polyphaga (a protozoan) and Thermoactinomyces. Repeated sampling and analysis for airborne bacteria and fungi in affected offices demonstrated only very low concentrations [<50 colony forming units (cfu/m^3)] of viable microorganisms.

These analytical results suggested that the agent causing illness was nonviable. Alternatively, the agent may have been viable but not culturable because of cell damage during sample collection or because unsuitable media was used during sampling and laboratory analysis. The agent causing this outbreak of illness may also have been a endotoxin (from bacteria) or a mycotoxin (from fungi).

AIR SAMPLING FOR MICROORGANISMS

One of the best reviews of the principles of sampling and analysis for microorganisms is found in the ACGIH Air Sampling Instruments manual [23]. Excellent reviews of sampling and analytical procedures especially applicable to indoor air quality studies have recently been published [24,25].

Even with excellent epidemiological and medical input, it is usually difficult to determine cause and effect relationships between analytical data and adverse health effects. For example, in some of the best documented hypersensitivity lung illness outbreaks in large office buildings, a causal relationship between environmental findings and disease could not be established [4,17]. In Building E, described earlier, several microbial amplification sites were present in offices but a clear association between any microbial isolate and human disease could not be established [16]. Building C provided an opportunity to associate the isolation of an environmental agent (Pseudomonas aeruginosa) from an amplifier (a portable humidifier) with human illness.

Without medical or epidemiological data indicating a plausible relationship between adverse health effects and a microbiologic agent, the interpretation of air sampling and analytical data becomes very difficult. For example, in Building B, it was clear that the porous manmade insulation in air handling units was an amplifier for fungi. An elevated concentration of fungi was detected in office air when the insulation was disturbed mechanically (Table 2). Building related illness, however, has not been documented among occupants. Therefore, great care should be exercised when hypothesizing cause and effect relationships between bioaerosol data and human illness.

Air sampling for microorganisms is governed by the principles that affect the collection of any particulate aerosol [25]. The collection of microorganisms is, however, complicated by the viability and stability of the sample. Three basic types of air samplers are used to collect most bioaerosols, namely impactors, impingers, and filter devices.

The proper choice of bioaerosol sampling instrument depends upon the preknowledge of the type of agent, its concentration, and properties. Impingement devices are useful when high concentrations of microorganisms are expected. Culture plate impactors including multiple and single stage sieve instruments and slit-to-agar instruments are useful in most office environments where lower concentrations of microorganisms are expected. Filter cassette samplers are useful probes for microorganisms resistant to desiccation.

At present, clear numerical guidelines and dose/response relationships between airborne concentrations of certain microorganisms and adverse health effects do not exist. The investigator must therefore consider the following parameters when interpreting bioaerosol sampling and analytical data:

1. The kinds and relative frequencies of microorganisms or microbial agents present in the indoor environment and in the outdoor air (rank order assessment).

2. Medical or laboratory evidence that an infection or allergy is caused by a kind or kinds of microorganism. If evidence indicates that illness is caused by a specific microorganism (for example, Legionella pneumophila serogroup 1), then the objective of the sampling and the interpretation of analytical data can be precise.

3. Indoor/outdoor concentration ratios. A high ratio for saprophytic fungi suggests that an indoor amplifier is present. A high ratio for gram positive bacteria, such as Staphylococcus, also suggests the occurrence of indoor amplifiers. In the case of Staphylococcus in Building D, the amplifiers are the office occupants themselves.

4. Concept of reservoir, amplifier, and disseminator [9]. In indoor environments, microbial agents may accumulate in reservoirs, and their populations may be amplified in niches such as wet porous insulation. However, for an illness to develop, a sufficient amount of the microbial agent must be transported to the breathing zone of a susceptible occupant.

SPECIAL TECHNICAL ISSUES

The investigator dealing with microbial agents in buildings must be aware of an increasing number of niches that may function as reservoirs, amplifiers, and disseminators for microbial agents.

The extensive use of porous insulation especially within the HVAC system of buildings (for example, Buildings A and B, Table 1) can provide niches where nutrients for microbial growth accumulate. Porous insulation located downstream of cooling deck coils is often moist during the air-conditioning season and as such becomes an ideal niche for the luxuriant growth of microorganisms.

The cold water present in some water spray systems has been shown to promote the amplification of some psychrophilic (optimal growth at low temperatures) _Cytophaga_ species [20] which have been putatively associated with an outbreak of hypersensitivity pneumonitis. By contrast, rubber washers and other construction materials in moderately warm hot water services can be amplification sites (biofilms) for bacteria such as _Legionella_.

The possibility that microbial volatiles are produced from HVAC system components and building construction materials should not be overlooked. Specific fungi such as _Aspergillus versicolor_ which is a common isolate in indoor environments are known to produce odorous volatiles which may affect ones perception of the acceptability of indoor air [26]. Proteinaceous additives present in floor covering materials are known to provide nutrients for anaerobic _Clostridium_ species. Putrefactive fermentation products, including putrescine, can be released into the indoor environment from this unusual source [27].

THE LANDLORD AND TENANT

The needs, concerns, and viewpoint of the landlord and tenant are often quite different from that of the investigator. The landlord and tenant often focus on matters, such as verification on whether a microbiological problem actually exists, the need for prompt remedial action, and protection from litigation. Some extreme, but not uncommon examples follow which illustrate the needs, concerns, and viewpoints of the landlord and tenant.

The Settle Plate Sampler

A tenant in a large office building hired a consultant to monitor indoor air quality. Settle plates containing malt extract agar were used to sample for indoor fungi. Six _Cladosporium_ colonies and two

Sporobolomyces colonies were recovered from a plate left in an office that housed an occupant who reported occasional headaches at work. The report provided by the consultant contained a laboratory data sheet from a subcontractor which stated that when 7 to 10 colonies are found, occupants will have "illness" and that occupants "...should consult with their personal physicians."

After reading this report, occupants became alarmed. The tenant requested that the building owner initiate corrective actions to eliminate the health hazard. The owner was aware of some nonspecific annoyance and discomfort complaints such as headache and eye irritation (possibly suggesting a ventilation deficiency), but did not have any idea on how to approach the alleged microbiological problem in the tenant office.

Lessens that the landlord, tenant, and consultant should have learned include the following:

1. Fungi are normally present in interior environments. To be able to interpret microbial sampling results, there must be adequate controls (outdoor air samples) and a rationale (preferably medical) on which sampling strategy was based.

2. Settle plates are not acceptable samplers for bioaerosols [25].

3. There was no reason to conduct bioaerosol sampling in this building.

Organism Suggestive of Legionella

As part of a proactive environmental program, a building owner hired a consultant to evaluate air quality in a number of its office buildings. The consultant performed sampling for a number of air contaminants, including bacteria that might be present in the indoor air. Settle plates, sterile swabs, and the All Glass Impinger samplers were used to collect bacteria. Swab samples were streaked across media containing buffered charcoal yeast extract media (without antibiotics), and the consultant's report indicated that a bacterium "suggestive" of Legionella was present.

The landlord was impressed by the techniques used by the consultant, especially the All Glass Impinger sampler which allowed "...one to see into the sampler itself" and because the sampler was "...run for exactly 30 minutes."

The landlord, however, was perplexed over the conclusion that bacteria suggestive of Legionella were present indoors. Furthermore, the landlord was concerned about the increased risk of litigation if this information was withheld from tenants in the affected building.

Lessons that the landlord, tenant, and consultant should have learned include the following:

1. Environmental sampling for Legionella is generally justified only when medical or laboratory tests indicate that certain species and serotype(s) have caused disease in susceptible occupants.

2. Water samples (not air samples) are commonly collected to determine if species or serotype(s) that have caused disease are present.

3. A walkthrough visual examination of the building and its offices to detect microbial reservoirs, amplifiers, and disseminators is always advisable before air sampling.

Six-Month Bacterial Sampler

A contractor provided maintenance services for an office building. Filter cassette samplers were placed in air supply ducts, left for six months, and were then analyzed for viable bacteria. The report provided to the building owner stated that the number of bacteria found on the sampler was less than 500 which is acceptable for ventilation air in offices. Tenants in the building who for a number of years had reported complaints including sinus congestion and cough as well as influenza-like illness doubted the validity of these tests and threatened the owner with litigation unless corrective actions were taken. The owner's counsel countered that the bacterial sampling program provided by the maintenance contractor guaranteed that the air in the building was acceptable for office occupancy.

Lessons that the landlord, tenant, and consultant should have learned include the following:

1. Most viable bacteria with the exception of a few spores, will be killed by desiccation on the filter. Probably only those bacteria collected in the last few minutes of the sampler's operation would have remained viable and thus culturable. Therefore, analytical results are suspect and essentially worthless.

2. Viable bacteria are collected most efficiently by impactor or impingement samplers where vegetative cells are protected against desiccation.

3. There was no medical or laboratory evidence that occupant illnesses were due to bacterial agents. Occupant maladies may have been caused by other types of microbial agents (for example, viruses) or may have been a result of normal community-associated illness.

Use Of Biocide In Active HVAC System

The facilities engineer of a large office building read an article in an engineering magazine that stated Legionella pneumophila was a "common" organism in the indoor air. Occupants of the building had, especially during the winter and summer months, reported numerous complaints of eye and nasal irritation, headache, and fatigue. Remedial actions suggested in the magazine article included a recommendation that microbiocidal chemicals be injected into the ventilation airstream at a point just downstream from fans. The owner of the building was concerned about the technical accuracy of the magazine article.

Lessons that the landlord, tenant, and consultant should have learned include the following:

1. Legionella is not a common microorganism in indoor air.

2. Legionella species and serotypes may cause Pontiac fever and Legionnaires' disease. These building related illnesses are diagnosed by distinct symptoms and laboratory tests. Many biological and nonbiological agents may elicit non-specific symptoms such as mucous membrane irritation, headache, and fatigue.

3. Microbiocidal agents used in active HVAC systems may adversely affect human occupants of the building. Microbiocidal agents should be used in HVAC systems only under carefully controlled conditions.

OBJECTIVES OF CONFERENCE

A primary concern of affected tenants and landlords, as illustrated in the previous examples, is the correct interpretation of technical data. If an occupant problem relating to bioaerosols is real, both the building owner and tenant must know what types of corrective actions are cost effective and appropriate.

ASTM is an organization that has been concerned with developing guidelines and test procedures for measuring the properties of a wide variety of materials. A primary objective of this conference is to provide information on sampling, analytical procedures, and interpretation of data for 12 types of microbial agents (chapters that follow) that are commonly or uncommonly found in interior environments. It is anticipated that the building owner, tenant, and investigator will all benefit from the diverse opinions expressed during these proceedings.

REFERENCES

[1] Feeley, J. C., Sr. and P. R. Morey, "Two Consultants Views of Tomorrow," Conference on Biological Contaminants in Indoor Environments, ASTM,STP 1071, P. R. Morey, J. C. Feeley, Sr., J. A. Otten, eds., American Society for Testing and Materials, Philadelphia, 1990.

[2] Rich, S. and P. E. Waggoner, "Atmospheric Concentration of Cladosporium Spores," Science, 137, 962-965, 1962.

[3] Williams, R.E.O., O. M. Lidwell and A. Hirch, "The Bacterial Flora of the Air of Occupied Rooms," J. Hyg., 54, 512-523, 1956.

[4] Arnow, P. M., J. N. Fink, D. P. Schlueter, J. J. Barboriak, G. Mallison, S. E. Said, S. Martin, G. G. Unger, G. T. Scanlon and V. P. Kurup, "Early Detection of Hypersensitivity Pneumonitis in Office Workers," Am. J. Med., 64, 236-242, 1978.

[5] Edwards, J. H., "Microbial and Immunological Investigations and Remedial Action After an Outbreak of Humidifier Fever," Bri. J. Ind. Med., 37, 55-62, 1980.

[6] Burge, S., A. Hedge, S. Wilson, J. H. Bass and A. Robertson, "Sick Building Syndrome: A Study of 4373 Office Workers," Ann. Occup. Hyg., 31, 493-504, 1987.

[7] Brundage, J. F., R. Scott, W. M. Lednar, D. W. Smith and R. N. Miller, "Building-Associated Risk of Febrile Acute Respiratory Disease in Army Trainees," JAMA, 259, 2108-2112, 1988.

[8] Fanger, P. O., J. Lauridsen, P. Bluyssen and G. Clausen, "Air Pollution Sources in Offices and Assembly Halls, Quantified by the Olf Unit," Energy and Buildings, 12, 7-19, 1988.

[9] Morey, P. R. and J. C. Feeley, Sr., "Microbiological Aerosols Indoors," ASTM Stand. News, 16 (12), 54-58, 1988.

[10] Feeley, J. C., Sr., "Legionellosis: Risk Associated With Building Design," In: Architectural Design and Indoor Microbial Pollution, R. B. Kundsin, ed., Oxford Univ. Press, New York, pp. 218-227, 1988.

[11] du Moulin, G. C., K. D. Stottmeier, P. A. Pelletier, A. Y. Tsang and J. Headley-White, "Concentration of Mycobacterium avium by Hospital Hot Water Systems," JAMA, 260, 1599-1601, 1988.

[12] Moser, M. R., T. R. Bender, H. S. Margolis, G. R. Noble, A. P. Kendal and D. G. Ritter, "An Outbreak of Influenza Aboard a Commercial Airliner," Am. J. Epidemiol., 110, 1-6, 1979.

[13] Marrie, T. J., A. MacDonald, H. Durant, L. Yates and L. McCormick, "An Outbreak of Q Fever Probably Due to Contact With a Parturient Cat," Chest, 93, 98-103, 1988.

[14] Bartlett, P. C., L. A. Vonbehren, R. P. Tewari, R. J. Martin, L. Eagleton, M. J. Isaac and P. S. Kulkarni, "Bats in the Belfry: An Outbreak of Histoplasmosis," Am. J. Pub. Health, 72, 1369-1372, 1982.

[15] Morey, P. R., "Experience on the Contribution of Structure to Environmental Pollution," In Architectural Design and Indoor Microbial Pollution, R. B. Kundsin, ed., Oxford Univ. Press, New York, pp. 40-80, 1988.

[16] Hodgson, M. J., P. R. Morey, M. Attfield, W. Sorenson, J. N. Fink, W. W. Rhodes and G. S. Visvesvara, "Pulmonary Disease Associated With Cafeteria Flooding," Arch. Env. Health, 40, 96-101, 1985.
[17] Hodgson, M. J., P. R. Morey, J. S. Simon, T. D. Waters and J. N. Fink, "An Outbreak of Recurrent Acute and Chronic Hypersensitivity Pneumonitis in Office Workers," Am. J. Epidemiol., 125, 631-638, 1987.
[18] Jacobs, R. L., R. E. Thorner, J. R. Holcomb, L. A. Schwietz and F. O. Jacobs, "Hypersensitivity Pneumonitis Caused by Cladosporium in an Enclosed Hot-Tub Area," Ann. Intern. Med., 105, 204-206, 1986.
[19] Marinkovich, V. A. and A. Hill, "Hypersensitivity Alveolitis," JAMA, 231, 944-947, 1975.
[20] Flaherty, D. K., F. H. Deck, M. A. Hood, C. Liebert, F. Singleton, P. Winzenburger, K. Bishop, L. R. Smith, L. M. Bynum, and W. B. Witmer, "A Cytophaga Species Endotoxin as a Putative Agent of Occupation-Related Lung Disease," Infection and Immunity, 43, 213-216, 1984.
[21] Morey, P. R., "Microorganisms in Buildings and HVAC Systems: A Summary of 21 Environmental Studies," In Proceedings of IAQ '88, Engineering Solution to Indoor Air Problems, Atlanta, GA, ASHRAE, pp. 10-24, 1988.
[22] Morey, P. R. and B. A. Jenkins, "What Are Typical Concentrations of Fungi, Total Volatile Organic Compounds, and Nitrogen Dioxide in an Office Environment?" In Proceedings of IAQ '89, The Human Equation: Health and Comfort, Atlanta, GA, ASHRAE, pp. 67-71, 1989.
[23] Chatigny, M., "Sampling Airborne Microorganisms," In Air Sampling Instruments, 6th ed. Cincinnati, ACGIH, pp. E2-E9, 1983.
[24] Burge, H. A. and W. A. Solomon, "Sampling and Analysis of Biological Aerosols," Atm. Env., 21, 451-456, 1987.
[25] Burge, H. A., J. Feeley, K. Kreiss, D. Milton, P. Morey, J. Otten, K. Peterson, J. Tulis and R. Tyndall, Guidelines For the Assessment of Bioaerosols in the Indoor Environment, American Conference of Governmental Industrial Hygienists, Cincinnati, Ohio, 1989.
[26] Hyppel, A., "Finger Print of Mould Odor," Proc. 3rd Intern. Conf. On Indoor Air Quality and Climate, 3, 443-447, 1984.
[27] Karlsson, S., E. Banhidi, Z. G. Banhidi, and A-C. Albertsson, "Accumulation of Malodorous Amines and Polyamines Due to Clostridial Putrefaction Indoors," Proc. 3rd Intern. Conf. On Indoor Air Quality and Climate, 3, 287-294, 1984.

TABLE 1 -- Potential Microbial Contaminants in HVAC Systems That May Be Associated With the Perception of Poor Air Quality

Component	Nature of contamination
Outdoor air intake	Emissions from sanitary vents, cooling towers, and decaying leaves.
Dust cake on filters of air handling unit	Organic materials in filter offer a substrate for growth of microorganisms if relative humidity exceeds 70%.
Insulation in air handling units and fan coil units	The ventilation air is exposed to microbial amplification sites especially where insulation is near or downstream from cooling coils.
Stagnant water in air handling units	The ventilation air passes over water where microorganisms may amplify. The potential for emissions of odors from microorganisms is great.
Fan coil and induction units	Dust and debris accumulate in units. Microorganisms may amplify when moisture is available (for example during the air-conditioning season).

TABLE 2 -- Airborne Fungi in Three Buildings Where Quiescent or Aggressive Sampling was Performed

Building, Time of Year	Microorganisms In Outdoor Air (cfu/m^3)[ab]	Ratio Indoor/Outdoor Microbial Concentrations	
		Quiescent Sampling In Office[c]	Aggressive Sampling In Office[c]
A June[d]	1350 F	10	---
A June[e]	60 T	100	---
B October[f]	200 F	0.08	Up to 200
C November[g]	830 F	0.07	13.2

[a]cfu/m^3 means colony forming units per cubic meter of air.

[b]F = fungi; T = thermophilic actinomycetes.

[c]Quiescent sampling occurs during normal conditions in offices or in the HVAC system. Aggressive sampling occurs when interior furnishings and/or HVAC system components are disturbed or moved.

[d]*Cladosporium* dominates outdoor air; *Penicillium* dominates indoor air.

[e]Thermophilic actinomycetes dominate indoor bacteria.

[f]*Cladosporium* dominates air indoors and outdoors.

[g]*Cladosporium* dominates all air samples.

TABLE 3 -- Sampling For Bacteria in Office With Portable Humidifier (Building C)

Sample and Location	Humidifier	Total Bacterial Count[a]	Rank Order
Air Sample in Office	Off	30	Bacillus
Air Sample in Office	On	3500	Bacillus >Corynebacterium >Pseudomonas
Outdoor Air	---	150	Bacillus >Micrococcus >Pseudomonas
Water Sample From Reservoir	---	1200	Pseudomonas aeruginosa >Flavobacterium >Pseudomonas

[a]Colony forming units per cubic meter (air) or organisms per milliliter (water).

TABLE 4 -- Concentration of Gram Positive Bacteria in a Crowded Office (Building D)

Sample Site	cfu/m^{3}[a]	Rank Order
Office	67	Staphylococcus aureus - 69% Micrococcus luteus - 31%
Office	43	Staphylococcus epidermidis - 56% Micrococcus - 28% Staphylococcus saprophyticus - 16%
Outdoor Air	38	Micrococcus - 58% Bacillus - 42%
Outdoor Air	2	Micrococcus - 100%

[a]cfu/m^{3} means colony forming units per cubic meter of air.

John C. Hierholzer

VIRUSES, MYCOPLASMAS AS PATHOGENIC CONTAMINANTS IN INDOOR ENVIRONMENTS

REFERENCE: Hierholzer, John C., "Viruses, Mycoplasmas as Pathogenic Contaminants in Indoor Environments", Biological Contaminants In Indoor Environments, ASTM STP 1071, Philip R. Morey, James C. Feeley, Sr., James A. Otten, Editors, American Society for Testing and Materials, Philadelphia, 1990.

ABSTRACT: Viruses spread by the respiratory tract, via droplets and fomites, are diverse in their biological properties and are commonplace in all populations. The diseases caused by these viruses, which occupy 11 different virus families and total about 270 serotypes, involve mainly the respiratory, conjunctival, and gastrointestinal tissues. Viruses with lipoprotein envelopes are labile and die quickly upon drying in the environment; viruses without an envelope are more stable and survive longer in air and on surfaces. The two oropharyngeal mycoplasmas have survival properties similar to enveloped viruses. All of these organisms can be sampled in air by impelling onto agar surfaces and eluting into cell culture medium for growth. Good personal hygiene, handwashing, and low humidity will decrease virus contamination.

KEYWORDS: viral diseases, respiratory viruses, mycoplasmas

"Viruses are submicroscopic obligatory intracellular parasites that lack energy-generating enzyme systems for independent replication. A complete infectious virus particle, or virion, is composed of a genome of either RNA or DNA surrounded by a protein shell or membranous envelope which protects the genetic material from the environment and allows the virion to pass from one host cell to another. Outside of the cell viruses are inert, but might be considered 'live' when viral nucleic acid enters the cell and causes synthesis of virus-specific protein and nucleic acid. Thus viruses could be considered as either very simple microbes or complex chemicals."

[1; CRC Press, reprinted by permission]

Dr. Hierholzer is a research microbiologist with the Respiratory and Enteric Viruses Branch (G-17), Division of Viral and Rickettsial Diseases, Center for Infectious Diseases, Centers for Disease Control, 1600 Clifton Rd. N.E., Atlanta, GA 30333.

Viruses are ubiquitous microorganisms found in most species of life, from bacteria to elephants, and in most environments of life, from the Arctic to the Sahara. While elephants and deserts are surely important, we are concerned in this conference with viruses that can be spread from person to person in indoor environments. These include the respiratory viruses, which are the agents of the common cold, influenza, pneumonia, conjunctivitis, and generalized diseases of the respiratory and intestinal tracts. These viruses can be spread by aerosolized droplets, such as are produced by sneezing and talking, and by fomites, such as by touching surfaces like doorknobs and towels which have been inadvertently contaminated with a person's secretions. Thus, these viruses are spread with ease in the everyday environments of schools, stores, workplaces, airplanes, etc., and are major contributors to the infectious diseases contracted and spread within indoor environments. This Chapter will review these viruses, how their spread can be minimized in the environments in which we find them, and how they can be sampled and detected by relatively easy techniques. I will not discuss viruses which are spread by close contact with skin (e.g., warts and other papillomaviruses), by injection of blood (e.g., hepatitis B; AIDS), or by animal or insect bites (e.g., rabies; yellow fever and other arboviruses).

1.0. CHARACTERISTICS OF AGENTS

The viruses discussed here include all the traditional respiratory viruses, which cause diseases of the upper and lower respiratory tract and conjunctivitis and which are spread via secretions from these sites, plus several viruses that are spread by droplets and fomites but which cause non-respiratory illnesses [2,3]. These viruses afflict everyone, causing an average of 3 respiratory illnesses per person per year and millions of lost work days per year, at an enormous economic cost. Many of the viruses cause epidemics of eye disease, pneumonia, gastroenteritis, rash illness, or generalized disease. And many which cause severe but recoverable illness in normal, healthy persons cause extensive and fatal illness in persons already compromised by immunodeficiencies or by underlying diseases of the lungs, heart, or liver.

There are over 200 viruses that cause these respiratory-spread illnesses (Table 1). They fall into 11 major groups, or families. The number and diversity of viruses have discouraged the development of vaccines against many groups of viruses and have also made specific identification of virus infections difficult. However, it remains important to identify respiratory viruses because proper epidemiologic control and preventive measures can only be instituted when the causative agent is known. Each virus group has a different morphology when seen under the electron microscope. The morphology is responsible for gross antigenic structure, and also for other biological aspects of the viruses that are involved with viability, ease of spread, air sampling, and laboratory identification [1-3].

Viruses are divided primarily into RNA and DNA groups, and secondarily into enveloped and nonenveloped groups. In contrast to other forms of life, viruses contain only one type of nucleic acid -- never both, and their envelope, if they have one, is a baggy shroud of proteins or glycoproteins containing a high percentage of lipids. The envelope is so important because it is essential for infectivity of the virus, yet it is a fragile structure. In general, RNA viruses with an envelope are the least stable in the environment; while DNA viruses without an envelope are the most stable. Examples of RNA viruses which have an essential lipoprotein envelope are rubellavirus, coronaviruses, influenza viruses, the parainfluenzaviruses, mumps, measles, and respiratory syncytial virus (RSV).

RNA Viruses

Rubella virus causes the disease German measles, or rubella, and is included here because it is readily spread by droplets from the throats of infected children, and to a lesser extent young adults. The disease rubella is not a respiratory disease: congenital rubella is a malformation disease of the fetus in the first trimester of pregnancy; rubella is a febrile rash illness in school-age children, where it explodes through the schools; and it is a mild rash illness with joint involvement in adults. Rubella virus is now relegated to sporadic cases, because the vaccine introduced many years ago has been very successful in eliminating the threat of epidemics. The virus belongs to the family Togaviridae, and it contains a single-stranded RNA in a 30-nm wide core and an envelope with short surface projections just 5 or 6 nm long.

Coronavirus is a major agent of the common cold. It is highly pleomorphic -- all enveloped viruses are pleomorphic, that is, they have many shapes and sizes due to the nature of the baggy envelope. The virions are 80-160 nm in diameter, with little definitive structure. Like rubella virus, coronaviruses have projections protruding from the envelope. The projections, or peplomers, are 10 x 20 nm and are club-shaped, thus suggesting the name "Corona-virus" for the taxonomic group (Fig. 1). Like the myxoviruses, coronaviruses have a single stranded RNA wound up with a protein; the complex is called a ribonucleoprotein (RNP) core. The 2 serotypes of human coronavirus that have been studied the most cause between 15 and 35% of mild colds in a given year and community. Laboratory tests for coronavirus must have intact virus as antigen, because all of its antigenic activities are located in the projections in close association with the envelope.

Influenza virus has hemagglutinin and neuraminidase projections in its pleomorphic envelope. The projections are 4 nm wide x 14 nm long. The hemagglutinin projections react with human, chicken, and guinea pig red blood cells to produce a classical hemagglutination (HA) pattern in standard HA tests; the neuraminidase projections contain the enzyme neuraminidase which cleaves the sialic acid site where the virus attaches to a red blood cell during HA. The virus has helical symmetry and a classic RNP: the RNA is wound up with large protein subunits into a helical nucleocapsid with the RNA running down the middle of the helix. Influenza virus causes the disease influenza, which is distinguished from the other respiratory diseases by its intense muscle pains and lack of coryza or runny nose. Influenza virus is readily spread by airborne droplets and by fomites, but annual vaccines prepared for the elderly and other high-risk groups are usually effective in minimizing severe epidemics.

Parainfluenza virus is a large, pleomorphic enveloped virus with hemagglutinin and neuraminidase projections. It is twice as large as influenza virus, ranging up to 300 nm in rough diameter. It has a distinct RNP, which looks somewhat like a herring bone. The parainfluenzaviruses are another major group of respiratory pathogens; although only 5 serotypes strong, they cause severe colds, croup, bronchitis, and pneumonia in children and adults. Occasionally, they also cause conjunctivitis and diarrhea. Two other viruses in the family Paramyxoviridae are mumpsvirus and rubeola (measles) virus. They are intermediate in size between influenza and parainfluenza viruses, and are enveloped viruses with single-stranded RNA in RNP form. Mumps virus causes classical mumps, with pronounced swelling of the parotid gland and sometimes orchitis; rarely, central nervous system (CNS) symptoms occur. Rubeola virus causes measles (or red measles), which is a severe generalized infection of the eyes, throat, respiratory tract, CNS, and intestinal tract. Although no vaccines are currently available for the parainfluenzaviruses, epidemics of both mumps and measles are preventable by a diligent immunization program. The "MMR" vaccine (mumps/measles/rubella) has been particularly effective in reducing epidemics of all 3 viruses in this country.

RSV is the most pleomorphic of all the respiratory viruses. It occurs in many large, bizarre shapes and sizes. RSV contains single-stranded RNA in an RNP, and an envelope, which fit the definition of the family Paramyxoviridae. RSV has 8 proteins, of which 3 are major antigens that are measured in laboratory tests: the G glycoprotein, the F or fusion glycoprotein, and the N or nucleoprotein (Table 2). The importance of the protein functions given in the table are exemplified by the failure of early vaccines against RSV. These vaccines were crude, formalin-fixed, alum-precipitated preparations that elicited antibody to the G glycoprotein, but the antibody was not protective. Instead, it formed antigen-antibody complexes when the immunized children became naturally infected and thus added to their illness. We now know that antibody to the F glycoprotein is required for protection. To date, RSV is still a severe pathogen of the lower respiratory tract in infants and in geriatric persons, causing significant mortality in both groups [4].

Thus, the enveloped RNA viruses that are spread by respiratory routes constitute 4 virus families and, because of their envelopes, are unstable in aerosols. The nonenveloped RNA viruses, on the other hand, are highly stable in the environment. Examples of these are poliovirus and viruses causing aseptic meningitis.

The family Picornaviridae constitute the largest group of respiratory viruses. These are small viruses, 18-25 nm in diameter, with cubic icosahedral symmetry, single-stranded RNA, no ribonucleoprotein, no envelope, and no projections. The viruses usually have to be concentrated by ultrafiltration or ultracentrifugation to be seen under the electron microscope. As seen in Table 1, the Picornaviridae family contains the polioviruses, the Coxsackie A viruses, the Coxsackie B viruses, the echoviruses, several more recent enteroviruses, and the rhinoviruses. All of these cause respiratory illness.

The polioviruses cause poliomyelitis, a scourge of past generations that is now succumbing to very effective vaccines against the three serotypes. The coxsackie-, echo-, and entero-viruses cause conjunctivitis and diarrhea in addition to upper respiratory tract illness. Also, some cause outbreaks of aseptic meningitis during the summer months; some cause peculiar rash illnesses, heart disease, and myalgia; and many occasionally cause pneumonia. Of special interest are Enterovirus type 70 and Coxsackie A24, both of which cause a distinct syndrome called acute hemorrhagic conjunctivitis that occurs in large epidemics [5], Enterovirus type 71 that causes a paralytic disease (polioencephalitis) reminiscent of polio, and Enterovirus type 72 (hepatitis A virus) that causes infectious hepatitis. Rhinoviruses as a group cause only mild colds. In general, the term "common cold" applies to coronaviruses and rhinoviruses, and after that to mild infections by other viruses.

The last single-stranded RNA virus listed is the calicivirus, which is probably spread by throat secretions as well as by fecal contamination, and causes gastroenteritis. Because members of the family Caliciviridae are so difficult to grow in cell culture or to test for, little is known of their biology. They are involved in nursery outbreaks of diarrhea and can probably be transmitted as a nosocomial pathogen in hospital wards.

The reoviruses are tough, double-stranded RNA viruses, also without an envelope, which are twice the size of caliciviruses but which appear not to cause overt disease. The reoviruses are isolated frequently enough to warrant their inclusion in this list, but any infection they cause is probably mild or asymptomatic.

DNA Viruses

Parvoviruses are small nonenveloped viruses, 20-25 nm in diameter, with a single-stranded DNA. They are very stable viruses, and easily spread by aerosolized droplets. The human parvovirus B19 was recently found as the cause of Fifth Disease (erythema infectiosum), a mild rash illness common in elementary school children, and to a lesser extent arthropathy, aplastic crisis in patients with chronic hemolytic anemia, and fetal death [6].

The herpesviruses have cubic icosahedral symmetry of their capsid and an envelope which renders them large and pleomorphic. Herpes simplex virus causes intense sore throats in primary infections. Its double-stranded DNA core looks much like an adenovirus capsid, since both have the same symmetry, but it is surrounded by a loose envelope with short (10nm) projections. All members of the family Herpesviridae are potent causes of disease. Herpes simplex type 1 causes gingivostomatitis and extremely sore throats in children during their primary infection, and can remain in their bodies to cause cold sores or fever blisters the rest of their lives. Herpes simplex type 2 causes painful genital lesions which tend to be recurring, making this an important venereal disease. Genital herpes infection can possibly lead to cervical cancer in later years. Both herpes simplex viruses cause severe conjunctivitis, rash illnesses with painful blisters, and encephalitis, along with multitudinous other symptoms of generalized disease. In general, type 1 causes more ocular, oropharyngeal, and generalized disease, and type 2 causes more venereal disease.

Cytomegalovirus (CMV) is another herpesvirus, and causes congenital hepatosplenomegaly and microcephaly in the unborn and mononucleosis and generalized disease in children and adults. It is midway in size between herpes simplex and varicella viruses, and is the only herpesvirus that is shed in great amounts in the urine. Varicella-zoster (VZ) virus is the largest herpesvirus, ranging up to 200 nm across the enveloped particle, and has a most peculiar disease association: it causes varicella (or chickenpox) in children -- a severe generalized disease affecting all parts of the body -- and it causes herpes zoster (or shingles) in adults, which is usually manifested by painful lesions around the waist.

Another herpesvirus with diverse manifestations is the Epstein-Barr virus. This is a smaller virus, like CMV (~120 nm), and it causes infectious mononucleosis (kissing disease) in young adults; this illness is characterized by fever, lymphocytosis, and a lengthy duration of malaise. Epstein-Barr virus is also associated with Burkitt's lymphoma and nasopharyngeal carcinomas in certain populations. The last human herpesvirus is type 6, which was tentatively associated with chronic fatigue syndrome at one point, and is a cause of roseola infantum. All herpesviruses, in particular simplex type 1, are spread by droplets; some are also spread by fomites and contact.

Polyomaviruses constitute a small group of little understood non-enveloped, double-stranded DNA viruses, about 45 nm in diameter. The JC virus appears to be a mild respiratory pathogen in children, and it also causes a rare, fatal disease called progressive multifocal leukoencephalopathy in children and young adults. The BK virus causes ureteral stenosis and acute hemorrhagic cystitis in children and in immunocompromised patients. Both viruses are spread by aerosols.

Adenovirus is a nonenveloped DNA virus 70-90 nm in diameter, with cubic icosahedral symmetry. From each of the 12 corners, or vertices, there is a projection which bears many type-specific antigens. The projections are thin and long, up to 20 nm. The main capsid component (hexon) is group-specific and is used in many diagnostic tests because of its stability and broad specificity. There are 47 known serotypes, and many more intermediate strains which cause disease but are difficult to place taxonomically [7]. The many diseases associated with adenovirus infection are shown in Table 3. Since these viruses are so easily spread both by droplets and fomites, they are a major contributor to disease contracted in indoor environments.

The predominant human respiratory mycoplasmas are listed last in Table 1 because they are frequently isolated in virology laboratories, and are not discussed elsewhere in this conference. Mycoplasmas are not viruses, and taxonomically fit somewhere between bacteria and rickettsia. Their outer membranes contain a high lipid content, which renders them fragile in the environment, and their antigenic structure is notably complex. _M. pneumoniae_ causes severe colds, pneumonia, conjunctivitis, myalgia, and diarrhea in children and young adults, but is rarely fatal. _M. hominis_ causes conjunctivitis, cervicitis, and occasionally respiratory symptoms in all age groups [8]. Mycoplasmas are thought to act synergistically with bacteria or viruses in some respiratory and CNS infections. Both species are readily spread by droplets and fomites.

2.0. SOURCE OF AGENTS

Viruses are found throughout the world, both in man and in all forms of animal and plant life. This statement is stressed because some viruses which now infect man (e.g., enterovirus type 70 causing hemorrhagic conjunctivitis) appear to have mutated from domestic or wild animals at a discernible time and place [5]. Other viruses (*e.g.* smallpox, herpes) appear to have been plagues of man as far back as history is recorded. Ethnic populations in all geographic and climatic regions of the earth, are now subject to respiratory virus infection, although before the era of colonial expansion in the Middle Ages there were large island and subcontinent populations which had never before experienced illnesses like yellow fever, measles, and polio.

Transmission of Viruses

Transmission of viruses from one person to another via respiratory routes occurs along two lines (Table 4). The most frequent route is aerosolized droplets. In this route, a person recently-infected with a respiratory virus produces copious quantities of new virus particles in his eye, nose, and throat tissues and mucous membranes, and then he inadvertently expels these particles in the form of small droplets of virus-laden saliva when he talks or laughs. He expels much more virus in the form of larger droplets when he sneezes or coughs, and these droplets will travel further in the air due to the force of the sneeze or cough. All of the droplets are suspended in the air like an invisible mist. The length of time that droplets are infectious depends upon many factors, *viz.*, the size of the droplets, the temperature and humidity of the air, the stability of the virus (whether enveloped or not), and the quantity of virus.

Person-to-person transmission of viruses also occurs via fomites. This route of transmission is vastly underestimated, and few solid studies have been done to verify its extensiveness. Fomite transmission occurs whenever a recently-infected person sneezes or coughs on an object or surface, or transfers secretions from his body onto an object (generally by touching), and then an uninfected person transfers the secretions from the object or surface to his own mucous membranes (eyes, nose, mouth). The secretions and excretions transferred are from the eye, nose, mouth, vesicles or lesions, urine, and stool. The objects and surfaces contaminated are items like towels, pillowcases, drinking glasses, eating utensils, tabletops, doorknobs, sink faucets, and light switches. For these reasons, fomite transmission is a particularly nettlesome problem in hospitals, where nosocomial infections are frequent and are very difficult to control [2,9]. In clinics such as ophthalmology offices, the spread of virus from an infected patient via fomites to a noninfected one readily occurs also. Adenovirus type 8 has been isolated from eyedrops [10], type 8 viral antigens have been detected in various eye medications [10,11], and adenovirus type 37 antigens have been detected in eye medications, instruments, and air conditioning filters [12] in outbreaks of keratoconjunctivitis in ophthalmologists' offices.

Many viruses can be transmitted by close contact, such as from mother to baby, kissing, or by venereal routes. In these cases, it is not possible to determine the exact route of transmission since all routes are present. Obviously, good personal hygiene will dramatically reduce fomite transmission. As with droplets, the length of time that fomites contain infectious virus depends on ambient temperature and humidity, the stability of the virus, and the amount of virus present.

Epidemiology

The epidemiology of the respiratory-spread viruses is quite complex [2-9]. Some viruses are seasonal, occurring in the fall and winter months each year (RSV); some occur in the fall/winter months every 2 years (parainfluenza 1); some all the time (parainfluenza 3). Some occur in the spring in a cyclic fashion (parvovirus); and some are endemic and seldom cause peaks of infection (herpesviruses, adenoviruses). The enteroviruses occur all year round but are more prominent in the summer months. Many viruses peak in activity when the children return to school in the Fall, and when they return to school after the Christmas and Easter holidays (and exchange the viruses they picked up wherever they visited). Rhinoviruses follow this pattern, and typically 2 or 3 rhinovirus serotypes are circulating in any one community at any one time; when they run out of susceptible hosts, they move onto a new area, and other rhinoviruses then invade the community. Swimming pools can be a focal point for certain adenoviruses (types 3, 4, 7) causing pharyngoconjunctival fever in the summer months; these viruses are readily spread by oral and fecal contamination of the pool water, and will survive for days if the chlorine level is inadequate [3,13,14]. Mumps and measles viruses are able to cause discrete outbreaks of disease whenever the virus is introduced into a cohort of susceptible individuals. Influenza has sharp peaks of activity in the winter months each year because the viruses, notably influenza A, continually mutate and find susceptible populations.

None of the respiratory viruses have animal reservoirs, except possibly influenza virus. Studies have shown that influenzaviruses isolated from horses, pigs, and ducks possess genetic similarities to human strains which were isolated during the same time period [2].

3.0. HEALTH EFFECTS

The respiratory viruses cause colds which can spread to generalized illnesses with a variety of symptoms. In children, viruses will often cause moderate-to-severe colds, because they have smaller airways (which become filled with fluid or constricted by swelling) and have little prior immunologic experience (by which specific antibodies can be made quickly). In adults, these infections are the well-known upper respiratory illnesses experienced several times a year.

Most colds will remain minor nuisances if the patient does not have risk factors that impede his body's attempt to fight the virus. Risk factors -- lifestyle influences such as stress, smoking, improper diet, and lack of rest; and underlying illness such as heart disease, diabetes, and cirrhosis of the liver -- are very important because they weaken the body and thereby allow a cold to develop into a lower respiratory illness or more generalized infection. The most potent host risk factor is the immunocompromised state, whether it be from infections like AIDS and hepatitis, from cancers of the immune system, or from drug regimens used to treat disease. Thus, immunocompromised patients are at greatest risk of severe infection with just about any virus, and fatal infections are common in these patients.

Colds occur in a wide range of symptoms. These vary from runny nose, sore throat, fever, diarrhea, and malaise, to additional symptoms of follicular conjunctivitis (pink eye), otitis media, rashes, lymphadenopathy, headache, myalgias, and involvement of the central nervous system or abdominal organs. Many of the viruses cause distinct syndromes that have been described in Section 1.0: rubellavirus - congenital malformations in the unborn, German measles (rubella); influenzavirus - influenza; parainfluenzavirus - croup, pneumonia; mumpsvirus - mumps; rubeola virus - measles; RSV - pneumonia; poliovirus - polio; coxsackieviruses and enteroviruses - aseptic meningitis, Bornholm's disease; enterovirus 70 - acute hemorrhagic conjunctivitis; enterovirus 72 - infectious hepatitis; calicivirus - gastroenteritis; parvovirus B19 - Fifth Disease; herpes 1 - gingivostomatitis, fever blisters; VZ virus - chickenpox, shingles; Epstein-Barr virus - infectious mononucleosis; adenovirus - pharyngoconjunctival fever, acute respiratory disease, acute hemorrhagic cystitis, gastroenteritis.

The illnesses also range in severity, from subclinical to fatal. The percentage of illnesses that are subclinical varies widely among virus groups. Viruses that tend to be mild in children or young adults, like rubella, coronaviruses, echoviruses, rhinoviruses, parvovirus, and polyomaviruses, tend also to cause large percentages of subclinical infections. This is known from extensive serosurveys, in which routine blood samples from diverse populations are tested for different viruses. Often, by the age of 20, nearly every person has developed specific antibodies to these viruses despite not having had overt disease. Conversely, the more severe viruses have a lower rate of subclinical infection: parainfluenzaviruses, mumps, measles, polio, coxsackieviruses, hepatitis A, herpesviruses, adenoviruses. Influenza is an exception in that it is a serious disease of adults, particularly the elderly, but is very mild in children. RSV causes severe pneumonia in infants and the elderly, but is a mild cold in other age groups. For these two viruses, the subclinical infection rate is highest in the least affected age groups, and age is clearly a host risk factor in susceptibility to clinical illness.

4.0. CASE STUDY

In 1989, many outbreaks of mumps and measles occurred on college campuses in the U.S. [15,16]. At first, these were thought to represent infection due to antigenic variants of the original strains. Detailed laboratory analyses, however, have shown that neither virus has changed antigenically; rather, the outbreaks arose because sufficiently large cohorts of susceptible students were gathered in one place when the viruses were introduced onto the campuses. Some of the susceptible cohort was due to vaccine failures when the students were vaccinated some 10 years before. Another segment of the cohort consisted of students who had never been vaccinated because they came from parts of the country or foreign countries where vaccination was not mandatory for entry into schools. Both viruses are readily spread by droplets, as evidenced by their rapid transmission in dormitories and classrooms.

In the past, we have found adenoviruses in eye drops [10-12], on ophthalmologic instruments [11,12], and in air conditioning filters [12] in ophthalmologists' offices. In all cases, the viral contamination was the cause of a large outbreak of eye disease that could have been prevented by proper disinfection of surfaces (e.g., with 0.5% Chlorox), useage of unit eyedrops, handwashing, and cohorting of patients. We have isolated other adenoviruses from swimming pool water that was improperly chlorinated and filtered [13,14].

5.0. SAMPLING, COLLECTION, AND TRANSPORT

For air samples, a portable SAS Air Sampler (Surface Air System, PBI International, Milano, Italy) which impacts particles against Rodac plates containing trypticase soy agar (TSA) is an efficient and simple design. Exposure times can vary between 1 and 4 minutes, but usually the shorter time is sufficient and prevents an excessive accumulation of mold spores. Alternately, the respirable stage of the Andersen Sampler can be used. For both samplers, the agar plates should be fresh because moist plates will trap more virus particles and the moisture will protect them from disintegration. The plates should then be sealed with a flexible tape (e.g., vinyl electrician's tape) and immediately shipped on cold packs or wet ice to the laboratory. Upon arrival, the plates are opened and the agar surfaces sliced off with sterile scalpels into 6-7 ml of viral culture medium, minced by vigorous vortex mixing, and clarified by centrifugation at 1000xg for 30 min. at 4 C. The supernatant fluids are inoculated into multiple cell cultures at 1.0 ml/tube.

For samples from environmental surfaces, wipe the surfaces or fomites with sterile, cotton-tipped swabs pre-wet with phosphate-buffered saline (PBS) and place them in tubes of bacteriologic medium for transport to the laboratory. As with other specimens, the samples should be shipped cold as soon as possible. The samples are processed for viral cultures the same as clinical specimens.

For clinical samples, the type of specimen and the manner of collection are dependent on the laboratory methods anticipated. Ideally, specimens should be collected within 2 days of onset of symptoms, because most viruses are only shed in the initial stages of illness. (Exceptions are adenovirus types 8, 19, and 37 in keratoconjunctivitis in which the virus is shed from the eye for 14 days; mumpsvirus, which is shed from the parotid gland and saliva for up to 12 days; and various adenoviruses and picornaviruses which are shed in the stool for several weeks after onset.) Nasal swabs are the easiest specimens to collect for respiratory viruses and are also the best specimens (i.e., contain the most virus) for the majority of the viruses described here. For nasal swabs, urogenital calginate swabs are inserted into the nasal passages, gently rotated to absorb mucus and cells, and then vigorously twirled into 2 ml of transport medium (such as tryptose phosphate broth with 0.5% gelatin, veal heart infusion broth, or trypticase soy broth), preferably without antibiotics. Throat swabs can be obtained with cotton-tipped wooden applicator sticks that are rubbed against the posterior nasopharynx and then placed in the transport medium; the stick can easily be broken off to leave the cotton tip in the medium. Eye swabs are collected with cotton swabs in a manner similar to throat swabs. Nasopharyngeal aspirates are collected with a neonatal mucus extractor and mucus trap to which transport medium is added. Swabs or scrapings of vesicular lesions are likewise obtained carefully and placed in transport medium. Urine and stool specimens are collected as for any pathogen. Which specimen to collect can be determined from Table 4 and from sites exhibiting clinical symptoms.

Specimens should be placed on wet ice and transported to the laboratory for immediate testing. This is particularly critical for specimens for fluorescent antibody tests because the epithelial cells must remain intact for a reliable test result. If testing is not possible within 5 days after collection, the specimens should be frozen on dry ice and stored at -70 C until processed, although this will decrease the amount of viable virus in some cases.

When processing for viral isolation, the specimens are treated with antibiotics, clarified at 1000x*g* for 3 min. at 4 C to remove cell debris and bacteria, and inoculated onto appropriate cell cultures. The cultures should include a human epithelial line (e.g., HEpII), human embryonic lung diploid fibroblast cells (e.g., HLF), human lung mucoepidermoid cells (NCI-H292), and human rhabdomyosarcoma cells (RD) for the broadest coverage of viruses within practical limitations. The inoculum (0.5 ml/tube) is adsorbed to the cell monolayers for 1 hr at ambient temperature, and the cultures are then fed with maintenance medium and incubated at 36 C for several weeks. Details of the media required for different cells and different viruses can be obtained from reference works [2,3]. Mycoplasma cultures are inoculated with 1.0 ml of untreated specimen (many mycoplasmas are sensitive to antibiotics) and incubated semi-anaerobically for up to 4 weeks [8].

6.0. ANALYTICAL PROCEDURES

Identifying viruses utilizes their diverse biological properties. Classical virology requires that viruses be grown in cell cultures and typed with specific antisera, using tests such as hemagglutination-inhibition, neutralization, or complement-fixation. These are good procedures but time-consuming because the viruses have to be grown (4-8 wks.) and then grouped (1-3 wks.). At the same time, virus culture is highly sensitive, being capable of detecting a single virus particle if enough subpassages are made to ensure its replication to a recognizable point. For this reason, virus culture is the preferred method of detecting virus in air or on environmental surfaces.

In the last 10 years, many laboratories have studied ways to more quickly identify the viruses, by looking for them or their antigens directly in clinical specimens. There are good reasons for doing this: simplifying our diagnostic tests; achieving greater sensitivity and reliability and a shorter turn-around time; and minimizing laboratory contamination. These goals have led to the development of a variety of clever test schemes, including two which are actively used today. Indirect fluorescent antibody (IFA) and enzyme immunoassays (EIA) have been improved by utilizing pure and potent reagents, among them monoclonal antibodies which ensure specificity and single-source supply. For example, adenovirus in human epithelial cells shows nuclear fluorescence, which is distinct from the cytoplasmic fluorescence of RSV in the same cells. For these viruses and many others, the use of monoclonal antibodies gives brighter fluorescence and less background than polyclonal antisera. Preferred EIA formats are outlined in Fig. 2 (an all-monoclonal antibody test) and Fig. 3 (an all-polyclonal antibody test) [17,18]. IFA and EIA tests are not as sensitive as culture for detecting the low levels of viruses found in environmental samples.

Currently under study is a new test developed at the University of Turku, Finland, -- the time-resolved fluoroimmunoassay (TR-FIA) [19]. This is basically a quantifiable IFA test, and it can be set up as either a direct or indirect test. The principle is as follows: A fluorescent probe with a long life-time is excited with a short light pulse (1 µs) and the specific fluorescence is measured after a 400 µs delay time. During the delay, the short-lived background fluorescence caused by the biological materials and plastic disappears, and the remaining, long-lived specific fluorescence is then counted for 400 µs. After a further delay of 100 µs, the excitation pulse and the measurement cycle is repeated about 1000 times during a total counting time of one second. The fluorescent probe used is a europium chelate which has a long decay time of several hundred microseconds. The excitation wavelength is 340 nm and the emission wavelength is 613 nm. Fluorescence is measured in a single photon-counting fluorometer (1230 Arcus, LKB/Wallec, Stockholm) equipped with a xenon flash lamp.

In principle, the direct or one-incubation TR-FIA begins with monoclonal IgG purified from mouse ascitic fluid (or polyclonal IgG from rabbit or horse hyperimmune antiserum) diluted in carbonate buffer to coat the wells of polystyrene microtitration strips. After incubation and washing steps, specimens (such as nasopharyngeal aspirates, nasal swabs, or stool suspensions) are diluted at 1:3 or 1:10 in a special diluent and added to the proper wells in the strips. Then, 0.1 ml of Eu-chelate-labeled antibody is added to each well. (The labeling of monoclonal antibodies with Eu-chelate is accomplished with isothiocyanatephenyl-EDTA, and the europium is actually non-fluorescent in this form. After completion of the immunoreaction and the binding of the label on the solid phase, the europium ion is dissociated with enhancement solution, and it then fluoresces.) As for IFA and EIA, we prefer monoclonal antibodies for the TR-FIA to enhance specificity and to lessen the background signal.

Thus, many respiratory viruses can be identified in direct antigen tests by IFA, EIA, and TR-FIA [20-28], although the utility of these tests in environmental studies has not yet been shown. Other tests, such as unusual EIA formats, DNA probes, and polymerase chain reaction (PCR) DNA amplification, are also in use, and these will be described in general terms in the following sections.

Rubella virus infections are documented only by EIA serology on the patient's acute and convalescent serum samples. Consideration of the clinical picture of the patient and of epidemiological factors greatly aid in diagnosing rubella infections.

Coronavirus can be directly identified in respiratory specimens by EIA, in which specimen virus is adsorbed to the plates, and rabbit antiserum, anti-rabbit IgG-alkaline phosphatase, and p-nitrophenyl-phosphate substrate are reacted in that order [20]. The usual diagnosis of coronavirus infection is accomplished by serologic methods after the patient has recovered. The peplomers constitute the hemagglutinating component; and thus we have a hemagglutination-inhibition (HI) test for OC43 and an indirect HA (IHA) test for 229E which are fairly sensitive [2]. The intact virion is necessary for neutralization, as well as for complement-fixation antigen; however, these tests are not very sensitive. The most widely used serologic test for coronavirus is EIA [2,21].

Influenza and parainfluenza viruses are identified by direct detection using IFA, EIA, and TR-FIA [2,18,19,22-25]. Cell culture isolates of these viruses are readily identified by IFA, EIA, HI, or neutralization tests. Serological tests to determine infection in a patient are most commonly the HI test and EIA tests. Mumpsvirus is tested for the same as the parainfluenzaviruses, and also is easily grown in cell culture. Rubeola virus (measles) is difficult to grow, and infections are usually determined after the fact by serological means (HI, EIA) [2,3,9].

Respiratory syncytial virus (RSV), the severe pathogen of the lower respiratory tract in infants and the elderly, can be detected in nasal specimens by IFA, EIA, and TR-FIA, or grown with difficulty in human epithelial cells [2,3,18,19,23-28]. Suitable serologic assays for RSV are CF, IFA, and EIA [2,3,9,28].

Picornaviruses in general are identified only by suitable culture and neutralization tests. This is time-consuming and costly in terms of supplies. The polioviruses are identified this way, or with probes that can distinguish the 3 serotypes. Other probe tests are currently being explored to detect coxsackieviruses and echoviruses without having to culture them, but such tests are not yet available [5]. The rhinoviruses can be serotyped by neutralization, but are generally identified as a group by their properties of being stable in the presence of chloroform (as a lipid solvent) but labile at pH 3. Picornavirus infection is measured serologically by neutralization tests, which have been simplified into micro-neutralization tests carried out in microtiter plates. In one recent study, a micro-neutralization test for enterovirus 70, the main cause of acute hemorrhagic conjunctivitis (AHC), compared well with "macro-" tests in cell culture tubes [29].

Calicivirus is identified solely by electron microscopic (EM) observation of virions in stool samples. Electron microscopy and immune electron microscopy (IEM), in which stool suspensions are mixed with the patient's own convalescent serum, are valuable tools for all the gastroenteritis agents and for most other viruses as well. It is, however, a highly specialized endeavor, so is not applicable as a general diagnostic method. Reoviruses are readily identified by electron microscopy and serotyped by HI [2,3].

Parvovirus B19 infection is documented by EIA tests on the patient's acute serum sample. This test has been devised as a direct antigen detection test, as a test for early (IgM) antibody, and as a test for late (IgG) antibody [6,30]. Electron microscopy and DNA hybridization probes have also been used extensively, and PCR tests are currently being evaluated.

The herpesvirus group is readily visualized by EM, with the cubic icosahedral shape of the capsid being prominent. The virus is typed by EIA and sometimes neutralization tests, but careful consideration is also given to the clinical picture presented. Other than herpes simplex type 1, the herpesviruses are difficult to grow in cell cultures [2,3].

The adenoviruses are among the easier viruses to identify because they are unique in producing prodigious quantities of soluble antigens as they grow in cell culture, and these antigens possess many type-specific and group-specific properties that lend themselves to a wide variety of diagnostic tests [2,3,9-14,17,19,22,24]. CF, IFA, EIA, and TR-FIA tests are group-specific, such that all human adenoviruses will react to about the same extent in these tests. HI and neutralization tests are type-specific. All of these tests are useful also as serologic tests [3,10-14]. Additionally, restriction enzyme analyses have been extensively employed as a molecular epidemiology tool in tracing epidemic strains of virus and in attempts at genotypic classifications of strains [3,7,31].

Spot hybridization probes to detect adenovirus directly in clinical specimens are being evaluated. In a study by Stalhandske et al. [32], purified adenovirus 2 DNA or cloned adenovirus 41 G fragment (by BamH1 endonuclease) was labeled with ^{32}P by nick translation and used as probe in the hybridizations. The specimen was extensively treated to extract the nucleic acids, which were then spotted onto nitrocellulose paper for the hybridization reaction. After a series of washing and incubation steps, the radioactivity was detected by autoradiography overnight. In a survey of 40 stool specimens, 18 were found positive for adenovirus by radioimmunoassay and 16 were found positive by the adenovirus 2 spot hybridization test [32].

Another dot blot technique was described by Takiff et al. [33]; this differed from the previous studies in that it was a type-specific test using a portion of the genome common only to adenovirus types 40 and 41 (the enteric adenoviruses) rather than the portion of the genome (coding for hexon antigen) common to all adenoviruses. In this study, the probes were plasmid recombinants containing the EcoRI endonuclease fragments of type 41, purified by cleaving the DNA with restriction endonucleases, separating on agarose slab gels, excising and eluting the appropriate bands from the gel, and precipitating the DNA fragment in ethanol. The DNA probe was labeled with ^{32}P-dCTP by nick translation. As in the previous procedure, the specimen, either stool suspension or cell culture supernatant, was extensively treated to obtain denatured DNA which was then spotted onto nitrocellulose for hybridization. The filters were dried and developed by autoradiography. The results compared well with the detection of adenoviruses by electron microscopy, cell culture, or EIA: 91% of type 40- or 41-positive specimens by these other means were found positive by dot blots. There were no false negatives, but there were some false positives which were due to other adenoviruses. Another important finding in this study was that antigen detection by dot blot was enhanced considerably by a brief passaging of the specimen in cell culture.

Hybridization techniques have been improved by utilizing better nick translation reagents, as used above, and by labeling probes with biotin to avoid the problems associated with radioisotopes. One in situ hybridization study utilized biotin-labeled dUTP in an adenovirus type 5 probe to detect virus in nasopharyngeal smears on glass slides [Gomes et al. [34]. After the probe reaction, the slides were incubated, washed, air-dried, counter-stained with fast green, dehydrated, mounted under Canada balsam, and then examined by light microscopy. Positive cells appeared distinctly brown in sharp contrast to a background of green-staining negative cells. The test required about the same amount of time as an IFA test. The in situ hybridization results were perfectly correlated with the IFA results in this study, and all adenovirus serotypes were equally well detected. The test had the disadvantage of requiring well-made smears that contained intact cells, but had the advantages (over dot blot hybridization techniques) of being much more specific, more sensitive, and workable with all kinds of specimens (stool, lung smears, NPA smears, etc.). In addition, no radioisotopes were used.

The human mycoplasmas are best identified by culture in selective agar media, followed by growth inhibition tests with specific antisera [8]. Serologically, mycoplasma infection is determined by CF tests with specially-prepared antigens. Most laboratories find mycoplasma work difficult because of the highly cross-reacting lipid membranes.

7.0. INTERPRETATION

Against this backdrop of virus characteristics and identification, what does the recovery of viruses from air samples or from environmental surfaces mean? The non-enveloped viruses are stable in the laboratory, and likewise in the environment. Many picornaviruses, particularly polio and hepatitis A viruses, have been found in significant numbers (infectious particles) downwind from sewage treatment plants and from where rivers carrying treated sewage empty into oceans [35-38]. The problem of shellfish posing a health hazard by their mechanical concentration of hepatitis viruses is well known [2,3]. Polio and hepatitis viruses have even been shown to survive in bottled mineral water for several months at ambient temperature [39], and poliovirus is particularly resistant to normal handwashing [40].

Rhinoviruses can be transmitted more effectively via fomites than via droplets [41,42]. Adenoviruses are stable for days in improperly chlorinated swimming pool water [13,14], and survive on ophthalmologists' hands and instruments long enough to be transmitted to the next patient [10-12]. Even enveloped viruses, which are usually labile in the laboratory, can survive long enough in the air to successfully infect persons passing by; this is obvious from past rampant epidemics of rubella, influenza, mumps, measles, and other respiratory viruses [15,16,43-46]. Parainfluenzaviruses were shown to reside in the environment of an enclosed Antarctic research station, causing severe outbreaks of respiratory disease whenever the personnel operating the base were exchanged [47]. The concept of enhanced respiratory disease resulting from tightly enclosed buildings was later explored by Brundage et al. in army inductees [48].

Therefore, the recovery of viruses from the air means that these virus particles are there, intact, and capable of causing infection if they can invade a new host before they disintegrate from sunlight or dryness and if the host is immunologically susceptible. The viruses are protected only by the mucus- and cell-laden secretions in which they were expelled. For most viruses, only a single particle (virion) needs to be present on a sensitive tissue (e.g., mucus membrane of nose, epithelial surface of conjunctiva) to initiate infection. Fortunately, the survival time for most viruses in air is very short -- less than a minute in some cases -- so that this vulnerability to sun and low humidity protects us to a fair degree [3,9,38,41,45].

8.0. INTERVENTION MEASURES

While unusual measures, such as iodine-treated paper handkerchiefs [42], have been proposed, good personal hygiene and thoughtfulness are the best defenses against the respiratory-spread viruses. Simple washing of the hands by an ill person interrupts spreading virus onto a fomite surface; conversely, avoiding hand-to-eye or hand-to-mouth contact by a well individual prevents carrying a virus from a surface to a sensitive tissue. Covering one's mouth when sneezing and coughing will prevent a forceful expulsion of virus into the air and greatly minimize the amount of virus shed. Trying to avoid contact with a person just "coming down with a cold" is highly effective, but not usually possible. Virucidal aerosol sprays, such as have been advertised for home use in recent years, should be avoided because the distorted view of disinfection presented by their manufacturers and the chemical contamination resulting from their use might do more harm than good.

In addition to handwashing and general hygiene, which will reduce fomite transmission, a practical solution to indoor airborne contamination is the redesign of air flow in enclosed spaces, e.g., in airplanes and schools. For example, conditioned air could be routed to circulate upwards and then pass through filters or ultraviolet light before being cooled and recirculated. Techniques for reducing the spread of viruses in doctors' offices and eye clinics have been promulgated in the literature, and were reviewed in Sect.4.0.

9.0. SPECIAL TOPICS

Many viruses are detected directly in clinical specimens by IFA, EIA, and TR-FIA with good results. The most sensitive system for the respiratory-spread viruses, however, is growing them in appropriate cell cultures. Because this involves so much time and expense, research continues into faster and less expensive methods of detecting them. DNA probe tests are one method, but at this time their future is not good: they are no more sensitive or available than EIA tests, and they require more steps, time, harsh chemicals, and usually radioactive labels [32-34]. PCR is another new method with great potential because of its accuracy in augmenting the viral DNA in a specimen and its sensitivity in detecting the amplified DNA. PCR will probably become a major diagnostic tool as specific primers become available.

10.0. CONCLUSION

The viruses spread via the respiratory tract (droplets or fomites) are diverse in nature, occupying 11 different virus families and totaling about 270 serotypes in number. The diseases caused by these viruses are mostly respiratory or conjunctival, but many of the agents cause distinct syndromes or generalized infections. Viruses are classified into RNA- or DNA-containing viruses, and then into envelope- or nonenveloped viruses. Their biological properties are manifold and form the basis of the many antigen-detection and serologic tests that have been developed for them. Epidemiologic factors (*e.g.*, age, season) also help in identification. The oropharyngeal mycoplasmas that cause illness in man are only two in number, but one of them causes severe disease. Viruses and mycoplasmas can be obtained from air by impeller-driven samplers which concentrate the organisms against an agar surface; the agar can then be cut off into cell culture medium, vortexed, and the supernatant fluid inoculated onto cell cultures to grow the viruses. The best prevention strategy for viruses is to avoid them by good personal hygiene; good hygiene by ill persons also will reduce the amount of virus in the environment for others to encounter. Redesign of air flow in enclosed spaces could greatly help to minimize exposure, by circulating air away from people and filtering it before it is passed back to them.

REFERENCES

[1] Palmer, E.L. and Martin, M.L., *Electron Microscopy in Viral Diagnosis*, CRC Press, Boca Raton, Fla., 1988.

[2] Lennette, E.H., Halonen, P. and Murphy F.A., Eds., *Laboratory Diagnosis of Infectious Diseases: Principles and Practice, Vol.II, Viral, Rickettsial, and Chlamydial Diseases*, Springer-Verlag, NY, 1988.

[3] Schmidt, N.J. and Emmons, R.W., Eds., Diagnostic Procedures for Viral, Rickettsial and Chlamydial Infections, 6th. ed., American Public Health Assn., Washington DC, 1989.

[4] Hierholzer, J.C. and Tannock, G.A., "Respiratory syncytial virus: A review of the virus, its epidemiology, immune response and laboratory diagnosis," Australia Paediatrics Journal 22:77-82, 1986.

[5] Ishii, K., Uchida, Y., Miyamura, K. and Yamazaki, S., Eds., Acute Hemorrhagic Conjunctivitis: Etiology, Epidemiology and Clinical Manifestations, University of Tokyo Press, Tokyo, 1989.

[6] Anderson, L.J., "Role of parvovirus B19 in human disease," Pediatric Infectious Diseases Journal 6:711-718, 1987.

[7] Hierholzer, J.C., Wigand, R., Anderson, L.J., Adrian, T. and Gold, J.W., "Adenoviruses from patients with AIDS: A plethora of serotypes and a description of five new serotypes of subgenus D (types 43-47)," Journal of Infectious Diseases 158:804-813, 1988.

[8] Cassell, G.H., Kahane, I., Clyde, W.A., Dybvig, K., Howard, C.J., Pollack, J.D. and Tully, J.G., Eds., Proceedings of the 6th Congress of the International Organization for Mycoplasmology (IOM), Israel Journal of Medical Sciences 23:315-783, 1987.

[9] Anderson, L.J., Patriarca, P.A., Hierholzer, J.C. and Noble, G.R., "Viral respiratory illnesses," Medical Clinics of North America 67:1009-1030, 1983.

[10] Sprague, J.B., Hierholzer, J.C., Currier, R.W., Hattwick, M.A. and Smith M.D., "Epidemic keratoconjunctivitis: A severe industrial outbreak due to adenovirus type 8," New England Journal of Medicine 289:1341-1346, 1973.

[11] D'Angelo, L.J., Hierholzer, J.C., Holman, R.C. and Smith, J.D., "Epidemic keratoconjunctivitis caused by adenovirus type 8: Epidemiologic and laboratory aspects of a large outbreak," American Journal of Epidemiology 113:44-49, 1981.

[12] Keenlyside, R.A., Hierholzer, J.C. and D'Angelo, L.J., "Keratoconjunctivitis associated with adenovirus type 37: An extended outbreak in an ophthalmologist's office," Journal of Infectious Diseases 147:191-198, 1983.

[13] D'Angelo, L.J., Hierholzer, J.C., Keenlyside, R.A., Anderson, L.J. and Martone, W.J., "Pharyngoconjunctival fever caused by adenovirus type 4: Report of a swimming-pool related outbreak with recovery of virus from pool water," Journal of Infectious Diseases 140:42-47, 1979.

[14] Martone, W.J., Hierholzer, J.C., Keenlyside, R.A., Fraser, D.W., D'Angelo, L.J. and Winkler, W.G., "An outbreak of adenovirus type 3 disease at a private recreation center swimming pool," American Journal of Epidemiology 111:229-237, 1980.

[15] Centers for Disease Control, "Measles - United States, first 26 weeks, 1989," Morbidity & Mortality Weekly Report 38:863-872, 1989.

[16] Centers for Disease Control, "Mumps prevention," Morbidity & Mortality Weekly Report 38:388-400, 1989.

[17] Hierholzer, J.C., Johansson, K.H., Anderson, L.J., Tsou, C.J. and Halonen, P.E., "Comparison of monoclonal time-resolved fluoroimmunoassay with monoclonal capture-biotinylated detector enzyme immunoassay for adenovirus antigen detection," Journal of Clinical Microbiology 25:1662-1667, 1987.

[18] Hierholzer, J.C., Bingham, P.G., Coombs, R.A., Johansson, K.H., Anderson, L.J. and Halonen, P.E., "Comparison of monoclonal antibody time-resolved fluoroimmunoassay with monoclonal antibody capture-biotinylated detector enzyme immunoassay for respiratory syncytial virus and parainfluenza virus antigen detection," Journal of Clinical Microbiology 27:1243-1249, 1989.
[19] Halonen, P., Obert, G. and Hierholzer, J.C., "Direct detection of viral antigens in respiratory infections by immunoassays: A four year experience and new developments," In, Medical Virology IV, L.M. de la Maza and E.M. Peterson, Eds., pp. 65-83; Lawrence Erlbaum Associates, Inc., Hillsdale, New Jersey, 1985.
[20] Macnaughton, M.R., Flowers, D. and Isaacs, D., "Diagnosis of human coronavirus infections in children using enzyme-linked immunosorbent assay," Journal of Medical Virology 11:319-326, 1983.
[21] Macnaughton, M.R., "Occurrence and frequency of coronavirus infections in humans as determined by enzyme-linked immunosorbent assay," Infection and Immunity 38:419-423, 1982.
[22] Harmon, M.W. and Pawlik, K.M., "Ezyme immunoassay for direct detection of influenza type A and adenovirus antigens in clinical specimens," Journal of Clinical Microbiology 15:5-11, 1982.
[23] Anestad, G., Breivik, N. and Thoresen, T., "Rapid diagnosis of respiratory syncytial virus and influenza A virus infections by immunofluorescence: Experience with a simplified procedure for the preparation of cell smears from nasopharyngeal secretions," Acta Pathologica Microbiologica Immunologica Scandinavia, Section B 91:267-271, 1983.
[24] Sarkkinen, H.K., Halonen, P.E., Arstila, P.P. and Salmi, A.A., "Detection of respiratory syncytial, parainfluenza type 2, and adenovirus antigens by radioimmunoassay and enzyme immunoassay on nasopharyngeal specimens from children with acute respiratory disease," Journal of Clinical Microbiology 13:258-265, 1981.
[25] Grandien, M., Gardner, P.S., "Rapid viral diagnosis of acute respiratory infections: Comparison of ELISA and IFA for the detection of viral antigens in nasopharyngeal secretions," Journal of Clinical Microbiology 22:757-760, 1985.
[26] Anderson, L.J., Hierholzer, J.C., Bingham, P.G. and Stone, Y.O., "Microneutralization test for respiratory syncytial virus based on an enzyme immunoassay," Journal of Clinical Microbiology 22:1050-1052, 1985.
[27] Hendry, R.M., Godfrey, E., Anderson, L.J., Fernie, B.F. and McIntosh, K, "Quantification of respiratory syncytial virus polypeptides in nasal secretions by monoclonal antibodies," Journal of General Virology 66:1705-1714, 1985.
[28] Hornsleth, A., "Detection of respiratory syncytial virus IgG-subclass-specific antibodies: Variation according to age, and diagnostic value," Journal of Medical Virology 16:329-335, 1985.
[29] Hierholzer, J.C., Bingham, P.G., Coombs, R.A., Stone, Y.O. and Hatch, M.H., "Quantitation of enterovirus 70 antibody by microneutralization test and comparison with standard neutralization, hemagglutination inhibition, and complement fixation tests with different virus strains," Journal of Clinical Microbiology 19:826-830, 1984
[30] Anderson, L.J., Tsou, C., Parker, R.A., "Detection of antibodies and antigens of human parvovirus B19 by enzyme-linked immunosorbent assay." Journal of Clinical Microbiology 24:522-526, 1986.

[31] Adrian, T., Wadell, G., Hierholzer, J.C. and Wigand, R., "DNA restriction analysis of adenovirus prototypes 1 to 41," Archives of Virology 91:277-290, 1986.
[32] Stalhandske, P., Halonen, P., Pettersson, U., "Detection of adenovirus in stool specimens by nucleic acid spot hybridization," Journal of Medical Virology 16:213-218, 1985.
[33] Takiff, H.E., Yolken, R., Straus, S.E., "Detection of enteric adenoviruses by dot-blot hybridization using a molecularly-cloned DNA probe," Journal of Medical Virology 16:107-118, 1985.
[34] Gomes, S.A., Pereira, H.G., Russell, W.C., "In situ hybridization with biotinylated DNA probes: A rapid diagnostic test for adenovirus," Journal of Virological Methods 12:105-110, 1985.
[35] Hickey, J.L. and Reist, P.C., "Health significance of airborne microorganisms from wastewater treatment processes, Part I, Summary of investigations," Journal of Water Pollution Control Federation 47:2741-2751, 1975.
[36] Baylor, E.R., Peters, V. and Baylor, M.B., "Water-to-air transfer of virus," Science 197:763-764, 1977.
[37] Teltsch, B. and Katzenelson, E., "Airborne enteric bacteria and viruses from spray irrigation with wastewater," Applied and Environmental Microbiology 35:290-296, 1978.
[38] Akin, E.W., Jakubowski, W., Lucas, J.B. and Pahren, H.R., "Health hazards associated with wastewater effluents and sludge: Microbiological considerations, In, Proceedings of the Conference on Risk Assessment and Health Effects of Land Application of Municipal Wastewater and Sludges, B.P. Sagik and C.A. Sorber, Eds., The University of Texas, San Antonio, 1978.
[39] Biziagos, E., Passagot, J., Crance, J., DeLoince, R., "Long-term survival of hepatitis A virus and poliovirus type 1 in mineral water," Applied & Environmental Microbiology 54:2705-2710, 1988.
[40] Eggers, H.J., "Handwashing and horizontal spread of viruses," The Lancet ii:1452, 1989.
[41] Dick, E.C., Jennings, L.C., Mink, K.A., Wartgow, C.D. and Inhorn, S.L., "Aerosol transmission of rhinovirus colds," Journal of Infectious Diseases 156:442-448, 1987.
[42] Hayden, G.F., Hendley, J.O. and Gwaltney, J.M., "The effect of placebo and virucidal paper handkerchiefs on viral contamination of the hand and transmission of experimental rhinoviral infection," Journal of Infectious Diseases 152:403-407, 1985.
[43] McDermott, W., Ed., Conference on Airborne Infection, Bacteriological Reviews 25:173-382, 1961.
[44] Hers, J.F. and Winkler, K.C., Eds., Airborne Transmission and Airborne Infection, John Wiley & Sons, New York, 1973.
[45] Kundsin, R.B., Ed., Airborne Contagion, New York Academy of Sciences, NY, 353:1-328, 1980.
[46] Gerba, C.P. and Goyal, S.M., Methods in Environmental Virology, Microbiology Series 7:1-378, Marcel Dekker, Inc., NY, 1982.
[47] Parkinson, A.J., Muchmore, H.G., Scott, L.V., Kalmakoff, J. and Miles, J.A., "Parainfluenzavirus upper respiratory tract illnesses in partially immune adult human subjects: Study at an Antarctic station," American Journal of Epidemiology 110:753-763, 1979.
[48] Brundage, J.F., Scott, R.M., Lednar, W.M., Smith, D.W. and Miller, R.N., "Building-associated risk of febrile acute respiratory diseases in Army trainees," Journal of American Medical Association 259:2108-2112, 1988.

TABLE 1 -- Viruses and mycoplasmas spread by respiratory routes

Virus group	No. of serotypes
Rubella virus	1
Coronavirus B814, 229E, OC43, 692	4
Orthomyxovirus	
Influenza A, B, C	3
Paramyxovirus	
Parainfluenza 1, 2, 3, 4A, 4B	5
Mumps	1
Measles	1
Respiratory syncytial virus (RSV)	~3
Picornavirus	
Poliovirus 1, 2, 3	3
Coxsackievirus A (types 2, 4, 8-10, 16, 21, 24)	22
Coxsackievirus B (types 1-5)	6
Echovirus (types 1-7,9,11,12,16,19-22,25,29-31,33)	30
Enterovirus 68, 69, 70 (AHC), 71, 72 (hepatitis A)	5
Rhinovirus	~125
Calicivirus	1?
Reovirus 1, 2, 3	3
Parvovirus B19 (Fifth Disease Agent)	1
Herpesvirus	
Herpes hominis	2
Cytomegalovirus	1
Varicella-zoster	1
Epstein-Barr	1
Polyomavirus	2
Adenovirus	47
types 1-5,7,8,11,16,19,21,29,31,35,37,40,41 (gen'l)	
types 4, 7, 14, 21 (military)	
Mycoplasma pneumoniae; *M. hominis*	2
TOTAL	270

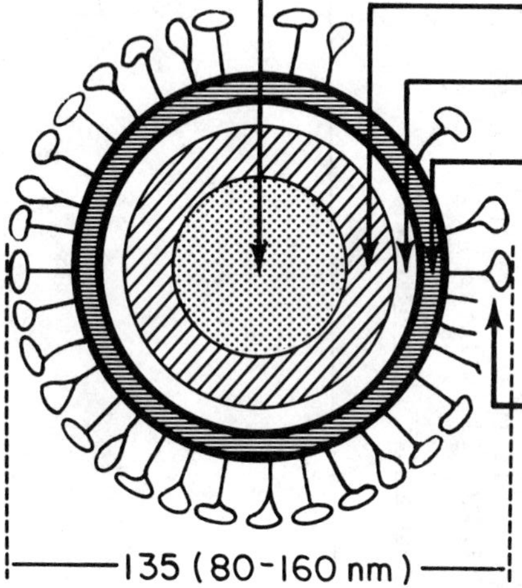

Scale: 1 cm = 20 mμ (nm)
or 1 mm = 2 nm

Structure	Width (nm)	Comp.*	MW	% of Viral Protein
Central cavity -RNP	: 46(35-50)	P	47,000	16
Inner shell	: 13(9-17)	P	30,000	26
Clear zone	: 6(4-8)	–	–	–
Outer shell	: 8(7-8)			
inner membrane layer		GP	60,000	23
outer membrane layer		LGP	191,000	13
Projections	10 wide	GP	104,000	8
	15-20 long	GP	15,000	14

*
P = Protein (polypeptide)
GP = Glycoprotein
LGP = Lipoglycoprotein

135 (80-160 nm)

FIG. 1 -- Diagrammatic structure of a human coronavirus

TABLE 2 -- Structural proteins of respiratory syncytial virus

Protein	Glycosylation	Molecular wgt. (kilodaltons)	Location	Function
L[a]	--	170-200	internal	RNA polymerase?
G	++	80-94	envelope	cell attachment?
F	++	66-74 43-59(F1);19-26(F2)	envelope	virus entry; cell fusion; neutralization; protection
N	--	41-44	internal	nucleocapsid protein
P	--	31-37	internal	phosphoprotein associated with N
M	--	27-29	envelope	matrix protein?
22K[a]	--	22-26	surface	unknown
1C	--	11-18	nonstruct.	unknown
1B	--	11-17	nonstruct.	unknown
1A[a]	++	9-10	unknown	unknown

[a] Minor component of virion. Internal proteins are part of the nucleocapsid structure; envelope proteins are surface antigens; nonstructural proteins are synthesized for virus replication.

TABLE 3 -- Adenovirus infections and serotypes involved

Syndrome	Signs and symptoms: Most frequent	Signs and symptoms: Less frequent	Ad serotypes usu. involved
Upper respiratory illness	coryza,pharyngitis, fever,tonsillitis	rash,otitis media, gastroenteritis	1,2,3,5,7,21
Lower respiratory illness	bronchitis,pneumonia, fever,coryza,cough	bronchiolitis	3,4,7,21,35
Pertussis syndrome	paroxysmal cough, vomiting,cyanosis	fever,URI	1,2,3,5
Acute respiratory disease	tracheobronchitis, fever,myalgia,coryza	pneumonia	4,7,14,21
Pharyngoconjunjunctival fever	pharyngitis,fever, conjunctivitis	coryza,headache, diarrhea,rash,nodes	3,4,7,11,16
Epidemic keratoconjunctivitis	keratitis,headache, preauricular nodes	coryza,pharyngitis, gastroenteritis	8,19,37
Acute hemorrhagic conjunctivitis	chemosis,follicles, subconjunc. hemorrhages	preauricular nodes, fever,coryza	11,15,37
Cystitis	cystitis (often hemorrhagic)	fever,pharyngitis	11,21,35
Immunocompromised host disease	diarrhea,rash, URI,pneumonia	hepatitis,cystitis, otitis media	11,29,34,35
Infant gastroenteritis	diarrhea,fever	nausea,vomiting mild URI	2,31,40,41
Neurologic disease	meningitis	encephalitis, Reye syndrome	3,7,32
Venereal disease	ulcerative genital lesions	cervicitis, urethritis	2,18,19,31,37

TABLE 4 -- Characteristics of viruses spread by respiratory routes[a]

Virus Family	Nucl. acid	Envel-ope	Median size (nm)	Virus or group	Virus present in E	N	T	L	U	S	Spread by drop.	fom.
Toga-viridae	ssRNA	+	~60	Rubella	-	±	+	-	±	±	+	-
Corona-viridae	ssRNA	+	~100	Coronavirus	-	+	+	-	-	-	+	+
Ortho-myxoviridae	ssRNA	+	~100	Influenza A,C	-	+	+	-	-	-	+	+
			~100	Influenza B	+	+	+	-	-	-	+	+
Para-myxoviridae	ssRNA	+	~250	Parainfluenza	±	+	+	-	-	-	+	+
			~170	Mumps	-	-	+	-	±	-	+	-
			~150	Measles	+	+	+	+	+	+	+	+
			~225	Resp.Syncytial	±	+	+	-	-	-	+	+
Picorna-viridae	ssRNA	-	26	Polio	±	+	+	-	±	+	+	+
			26	Coxsackievirus	±	+	+	±	±	+	+	+
			26	Echovirus	±	+	+	-	±	+	+	+
			26	Enterovirus	±	±	+	±	-	+	+	+
			27	Hepatitis A	-	-	-	-	±	+	±	+
			31	Rhinovirus	±	+	+	-	-	-	+	+
Calici-viridae	ssRNA	-	30	Calicivirus	-	?	?	-	-	+	?	+
Reo-viridae	dsRNA	-	75	Reovirus	-	-	+	-	±	+	+	±
Parvo-viridae	ssDNA	-	22	Parvovirus	-	+	+	-	+	-	+	+
Herpes-viridae	dsDNA	+	~160	Herpes simplex	+	+	+	+	±	-	+	+
			~120	Cytomegalovirus	+	+	+	-	+	+	+	-
			~180	Varicella-zost.	+	+	+	+	-	-	+	+
			~120	Epstein-Barr	-	-	+	-	-	-	+	+
Papova-viridae	dsDNA	-	45	Polyomavirus	-	+	+	-	+	-	+	?
Adeno-viridae	dsDNA	-	82	Adenovirus	+	+	+	+	+	+	+	+

[a] Virus Families have the suffix "-idae". Nucleic acid type is single-stranded (ss) RNA or DNA or double-stranded (ds) RNA or DNA. Viruses can have an envelope that is essential for infectivity (+) or not have an envelope (-). The median diameter of the virion is given in nanometers; enveloped viruses are pleomorphic (~). Virus may be present in (and therefore spread from) the eye (E), nose (N), throat (T), lesions (L), urine (U), or stool (S), which are the common sources of secretions and excretions involved in droplet and fomite transmission; virus present in internal tissues, cerebrospinal fluid, blood, etc., is not indicated here.

FIG. 2 -- The all-monoclonal capture EIA test [18]

1. Solid phase is a 96-well flat-bottom polystyrene microtiter plate (e.g., Immulon-2, Dynatech Laboratories, Alexandria, VA).

2. Add 75 µl/well of specific monoclonal antibody (MAb) (as capture antibody) as purified IgG diluted in carbonate buffer, pH 9.6; incubate overnight at 4 C; wash 3x with 0.01 M PBS, pH 7.2, containing 0.15% Tween-20.

3. Add 75 µl/well of specimen (NPA, etc.) at a low dilution (e.g., 1:5, 1:10, 1:20) in wash buffer containing 0.5% gelatin; incubate 1.5 hr at 37 C; wash 3x as in step 2.

4. Add 75 µl/well of same or different MAb-biotin (as biotinylated detector antibody), diluted in 0.01 M PBS, pH 7.2, with 0.5% gelatin, 0.15% Tween-20, and 2% normal goat serum; incubate 1 hr at 37 C; wash 3x as in step 2.

5. Add 75 µl/well of streptavidin/peroxidase (as conjugate), diluted in the diluent used for step 4; incubate 10 min. at ambient temperature; wash 6x as in step 2.

6. Add 125 µl/well of substrate system: 0.1 mg/ml of TMB in 2% DMSO in 0.1 M acetate/citrate buffer, pH 5.5, with 0.005% hydrogen peroxide added when used; incubate 20 min. at ambient temperature; stop color reaction with 2 M sulfuric acid; read at 450 nm in an automated EIA reader. [TMB is 3,3',5,5'-tetramethyl-benzidine.]

FIG. 3 -- The polyclonal capture EIA test [18]

1. Solid phase is a 96-well microtiter plate (see Table 5, step 1).

2. Add 75 µl/well of polyclonal antiserum as purified IgG (see Table 5, step 2).

3. Add 75 µl/well of specimen (NPA, etc.) at a low dilution (see Table 5, step 3).

4. Add 75 µl/well of a different species antiserum (as detector antibody), diluted in 0.01 M PBS, pH 7.2, with 0.5% gelatin, 0.15% Tween-20, and 2% normal species serum; incubate 1 hr at 37 C; wash 3x as in step 2.

5. Add 75 µl/well of anti-species IgG/peroxidase (as conjugate) in the diluent used for step 4; incubate 1 hr at 37 C; wash 6x as in step 2.

6. Add 125 µl/well of the substrate system: 0.1 mg/ml of TMB in 2% DMSO in 0.1 M acetate/citrate buffer, pH 5.5, with 0.005% peroxide added when used; incubate 20 min. at ambient temperature; stop color reaction with 2 M sulfuric acid; read at 450 nm.

DISCUSSION

If a viral probe were going to be conducted using impaction devices, what is the best way to ship the agar plates (Andersen, Rodac) from the field to the laboratory once sampling is complete?

CLOSURE

Ice packs are definitely the way to ship impacted petri plates. Freezing them on dry ice would preserve some viruses a little better, but would make it difficult to render the agar surface into a useable specimen. With overnight delivery, plates and ice packs can be sent easily and received for processing in the laboratory the following day. In fact, we have done this with Rodac plates several times, and have recovered adenoviruses, parainfluenzaviruses, echoviruses and other picornaviruses, respiratory syncytial virus, and various mycoplasma species. The only problem we have encountered is fungi in the viral cultures; this was overcome by filtering the contaminated cultures through disposable 0.22 μ filters before subpassaging.

DISCUSSION

Should viruses be sampled by wet collection systems (impinger/impactor) or by dry collection systems (filter cassette)? Should different collection methods be used for different families, genera, or serotypes?

CLOSURE

Viruses should be sampled by wet systems to maintain their viability. Dried viruses are dead viruses, and until rapid antigen tests become more available, viruses must be kept "alive" to culture them. Wet systems are useable for all virus families and genera.

DISCUSSION

Can parvovirus B19 be spread by aerosol in medical clinics? What type of engineering controls are appropriate to reduce the risk of aerosol transmission of viral diseases in outpatient clinics, general offices, and schools?

CLOSURE

B19 is a nosocomial pathogen, spread both by aerosols and fomites (H. Faden et al., Journal of Infectious Diseases 161:354-355, 1990). Epidemiologic data shows efficient spread within elementary school classrooms and within family units. Engineering design of the flow of conditioned air from the floor upwards will minimize the risk of infections by aerosolized viruses.

DISCUSSION

You mentioned that herpes simplex 1 and 2 could be infectious from surfaces and aerosols. Do you have an approximate number of infections that have occurred by these forms of transmission? How big a problem could this be in a crowded business office, say, with some individuals that are infected with herpes and who practice poor personal hygiene? Could you identify this type of problem by swabbing surfaces?

CLOSURE

Because herpes 1 infections are so common in children, who are exposed from multiple routes of transmission, it is difficult to know the percentage of infections that result from fomites vs. aerosols. Herpes 2 is spread predominantly by sexual contact. Herpetic lesions of either type which are active (and therefore contain high titers of virus) are very painful to the touch, and this reduces the chance of fomite spread; on the other hand, it is possible to contaminate a surface with lesion fluid via one's fingers without actually touching the lesion. For this reason, good personal hygiene is very important in office and home settings. If such a problem is suspected, surfaces such as desks and light switches can easily be swabbed with a moist cotton applicator. The chances of recovering herpes viruses from these swabs depends heavily on how much virus was initially present and how recently it was deposited (that is, you can recover virus so long as it did not "dry" out). Probably, the very act of swabbing office surfaces would serve to remind those present to be careful!

DISCUSSION

Has anyone demonstrated that good filtration and ventilation in classrooms decrease disease rates?

CLOSURE

Studies during World War II showed that enclosed, crowded barracks were prime targets for outbreaks of adenoviral and influenzal disease, and that these outbreaks could be ameliorated with adequate ventilation. Open windows in elementary school classrooms serve the same purpose. More recently, a study by Brundage et al. [48] showed increased risk of aerosol-transmitted infections in tightly enclosed buildings and a reduced risk in older, more open buildings.

DISCUSSION

Speaking of studies of ARD in Army barracks, especially the studies you just referred to showing increased risk in tighter buildings, what controls are feasible in such buildings if transmission is airborne rather than by fomites?

CLOSURE

Outside air can be mixed with recirculated air before it is conditioned, and this will dilute the airborne viruses generated by persons in the building. In seasons of obvious outbreaks of respiratory disease, the volume of outside air can be increased to effect a greater dilution. Also, the recirculated air can be passed through ultraviolet light, which is very effective against viruses.

DISCUSSION

With regards to epidemiology, how significant is the dormant characteristic of viruses when attempting to find sources and etiology?

CLOSURE

Viruses are never dormant in the biological sense of the word. Some can become latent in an infected person by retreating to nerve cells (e.g., herpes viruses), lymphoid tissue (e.g., adenoviruses), etc. They are not transmissible from the latent state. Rather, viruses are always spread by person-to-person or person-surface-person transmission. Therefore, the person who recently became infected serves as a potent source of virus for the next several days.

DISCUSSION

Could you describe the proper sampling and sample transport methods and what are the caveats of sampling for viruses?

CLOSURE

I believe that impactor samplers work well with 1, 2, or 3-minute exposure of the agar plates. The plates should be immediately sealed with tape and shipped with cold packs overnight to the laboratory. This prevents drying and deterioration and reduces bacterial and fungal growth on the plates.

DISCUSSION

What effect does the humidity level in room air have on the collection of airborne viruses?

CLOSURE

Humidity has a great deal to do with survival of viruses in air, and therefore with collection of airborne viruses. The air probably should be sampled longer (e.g., 4 or 5 minutes) in dry air because any viruses present will lose their infectivity quicker, and therefore the longer sampling time will increase the chances of picking up virus.

DISCUSSION

Assuming that herpes virus type 6 is the cause of chronic fatigue syndrome, can this disease be transmitted in air? How long do Epstein-Barr virus and herpes type 6 virus remain viable in air?

CLOSURE

Current data suggest that herpes 6 is not the cause of chronic fatigue syndrome, but is associated with a childhood rash illness called roseola infantum. Anecdotal information suggests that both herpes 6 and Epstein-Barr virus (EBV) are highly labile in air, and are spread predominantly by direct contact with secretions and only to a lesser extent by aerosols and fomites.

DISCUSSION

Can routine air sampling for viruses in general be helpful in surveillance of hospital environments or around infectious waste incinerators, etc.?

CLOSURE

Here you have a cost vs. benefit problem. I believe that routine virus sampling, with the necessary culturing and identification work, is too costly to be done on an ongoing basis. Large hospitals have "Infection Control Officers" who monitor suspicious outbreaks of a virus within a section of the hospital (nosocomial infections). It is routine practice in such situations to institute rigorous handwashing, mask and gown changes, disinfection of surfaces, and cohorting or isolation of patients to contain the outbreak. It is also feasible to sample air and surfaces in the affected parts of the hospital, including infectious waste incinerators, in an effort to trace the source of the outbreak. Environmental sampling would, of course, have to be done before the general disinfection steps were carried out.

DISCUSSION

What are the predominant viruses found in an office setting, and what is a good means to sample for these viruses from the air?

CLOSURE

In a typical office environment with middle-aged adults, influenza virus, rhinoviruses, and enteroviruses would predominate, because these are the agents of respiratory disease outbreaks among adults. They will likely be present only during the outbreaks, which are seasonal. Sampling methods have been discussed earlier.

DISCUSSION

What are the key aspects of sampling for aerosolized influenza and rhinovirus in a confined environment such as an airliner cabin?

CLOSURE

Air sampling can be done with impaction devices placed at air outlets and inlets of airliner cabins to determine the extent of viral contamination in the air and whether viruses are getting through the cooling system intact and being released back into the cabin. The same principles apply that have been discussed above, and the viral cultures done in the laboratory upon receipt of the agar plates can be optimized for these particular viruses.

DISCUSSION

Is antigen sampling an adequate way to look for viruses in air?

CLOSURE

Theoretically, yes. At the moment, however, we have little information on the durability of viral antigens in air. The adenovirus hexon antigen has been detected in ophthalmic solutions, air filters, etc. (see text, Sect. 4.0). Certainly, monoclonal antibodies used in EIA or TR-FIA tests are an excellent start for such studies.

DISCUSSION

Is air sampling for human viruses practical and useful?

CLOSURE

I believe it is practical when people are apparently getting sick in a building, and when there is epidemiologic evidence to confirm this. Often, the investigation will turn up a causative problem that the building managers did not know existed. Air systems and water systems have frequently been incriminated in such outbreaks.

F. Marc LaForce

BIOLOGICAL CONTAMINANTS IN INDOOR ENVIRONMENTS: GRAM POSITIVE BACTERIA WITH PARTICULAR EMPHASIS ON BACILLUS ANTHRACIS

REFERENCE: LaForce, F. Marc, "Biological Contaminants in Indoor Environments: Gram Positive Bacteria with Particular Emphasis on *Bacillus anthracis*", Biological Contaminants in Indoor Environments, ASTM STP 1071, Philip R. Morey, James C. Feeley, Sr., James A. Otten, Editors, American Society for Testing and Materials, Philadelphia, 1990.

ABSTRACT: Micrococci and staphylococci are commonly isolated when bacteriologic samples of room air are taken; however, their role as pathogens via the airborne route has not been established. Inhalation anthrax is the only well documented Gram positive lower respiratory infection that can follow airborne challenge. Pulmonary anthrax is now rare but at one time was an important hazard for textile workers exposed to goat hair imported from the Middle East. Aerosol challenge studies in experimental animals using anthrax spores have shown that particle size and total inoculum are important predictors of disease. Air sampling studies in goat hair processing plants showed that workers were regularly exposed to low levels of airborne anthrax but few cases occurred. The disease has all but disappeared, with the last United States case reported in 1977. There are no persuasive data suggesting that other airborne Gram positives are important pathogens and routine aerosol sampling for their presence is not recommended.

KEYWORDS: inhalation anthrax, air sampling, goat hair, aerosol challenges, Woolsorters' Disease.

INTRODUCTION

Gram positive bacteria are commonly isolated when bacteriologic studies of room air are done. Despite the frequent isolation of staphylococci and micrococci, their roles as airborne pathogens have not been established. In fact, experimental aerosol exposure studies have shown that high inoculums of these organisms are

Dr. LaForce is Chairman, Department of Medicine, The Genesee Hospital, and Professor of Medicine, University of Rochester School of Medicine and Dentistry, Rochester, NY 14607.

rapidly cleared by lung phagocytes. The single important exception to this general rule is respiratory infection due to B. anthracis. The clinical syndrome of pulmonary anthrax can follow the inhalation of anthrax spores and is probably the best studied bacterial pathogen which can be spread by the airborne route. Clinical and epidemiologic studies have accurately described Woolsorters' disease, the clinical hallmark of airborne anthrax. Experimental studies with anthrax spores first established the critical importance of particle and inoculum size in the pathogenesis of inhalation anthrax. Lastly, it was the first airborne disease which was controlled by decontaminating infected raw material.

CHARACTERIZATION OF B. ANTHRACIS

Bacteriology

Anthrax is a zoonotic infection which predominantly affects domesticated herbivores. Man is infected as a result of contact, inhalation or ingestion of B. anthracis usually through contact with an infected animal or their products. Cutaneous anthrax is the commonest clinical disease due to this organism. B. anthracis is a gram positive sporulating rod which is part of a large family of soil organisms which are for the most part non-pathogenic for man; B. anthracis is the major exception. The organism grows easily on laboratory mediums and can be recognized on bacteriologic plates by its shiny grayish appearance and a peculiar tendency to stand up in peaks when individual colonies are touched with a bacteriologic loop. This is a useful characteristic when screening bacteriologic plates for the presence of B. anthracis (1). A gamma bacteriophage is specific for anthrax and is useful for preliminary identification of this organism. Virulence of B. anthracis is dependent upon plasmid-mediated production of a three component exotoxin and an antiphagocytic ploy-D-glutamic acid capsule. The toxic proteins, collectively referred to as anthrax toxin, have been purified and cloned and consist of a protective antigen (PA), an edema factor (EF) and a lethal factor (LF).

Like all members of the Bacillus species B. anthracis sporulates. Spores encase the genome of B. anthracis in an insulating dehydrated vehicle that makes the organism ametabolic and resistant to a wide variety of lethal agents. The spore coat is made up of a keratin-like material which constitutes as much as 80 percent of the total protein of a spore. This protein is relatively impervious to chemicals and is responsible for the survival characteristics of spores. In the spore state B. anthracis can remain viable indefinitely. However, germination can cocur swiftly under the proper circumstances.

RESERVOIRS OF B. ANTHRACIS IN THE ENVIRONMENT

Persistence of Anthrax in the Soil

B. anthracis is the first organism whose life cycle was fully described and these studies brought a measure of early fame to

Robert Koch (2). He confirmed that in animals dying of anthrax he could find filiform bodies in their blood. Mice inoculated with the blood from infected animals died and he could recover the same filiform bacterial from mouse blood. However, an unresolved question was how the organism survived in order to repetitively cause disease in grazing animals. Koch chose to study this problem by following the growth of B. anthracis microsopically outside the animal. Using the sterile aqueous humour from beef eyes, he followed the multiplication of anthrax in vitro. He noted that after about fifteen hours of bacterial multiplication the box-car bacilli gave rise to round refractile structures, spores. He showed that these spores could revert into vegetative forms in fresh media and cause anthrax if injected into experimental animals. Koch clearly understood the meaning of this transformation and wrote "But the spores survive in a hardly believable manner and way. Neither years of dryness, nor existence in a putrescent fluid for months, nor repeated drying and moistening can destroy their power of germination. When these spores have formed, there is ample reason for the anthrax not disappearing for a long time in a certain region." Tissue from animals dying of anthrax can be heavily contaminated with anthrax spores which can survive indefinitely.

Importance of Imported Goat Hair as a Source of Anthrax Spores

In England prior to 1837 no specific disease had been associated with wool. From 1837 to 1847, after mohair from Asia Minor had been introduced as textile fibers, deaths among Bradford woolsorters began (3). In 1879, barely two years after Koch's description of the life cycle of anthrax, John Bell, a Bradford physician, bled a patient who was dying of Woolsorters' disease. He examined the blood microscopically and saw many filiform organisms. He inoculated infected human blood into experimental animals and was able to transmit the disease (3).

Bell's studies led to the enactment of the Bradford Rules which were specifically enacted to reduce exposure of workers to dusts from imported goat hair. Because of the continued occurrence of occasional cases, the Anthrax Investigation Board was established. Its work resulted in the first detailed microbiologic study of imported hair products. Over 14,000 specimens were studied bacteriologically, the results of which led to a sound classification of dangerous raw materials. Not surprisingly, goat hair from East India, Persia and Turkey led the list in positive isolations. In the United States a small industry used imported goat hair to manufacture a hair cloth which was used as an interlining of men's suits. This industry has been the most important source of airborne anthrax cases in the United States.

CLINICAL SEQUELAE OF AIRBORNE EXPOSURE

Woolsorters' Disease

Woolsorters' disease, or inhalation anthrax, is a necrotizing hemorrhagic mediastinal infection that is almost always fatal.

Infection follows inhalation of anthrax spores which are phagocytized by alveolar macrophages and transported to hilar and tracheobronchial nodes where germination and multiplication of bacilli occur. Since 1900 there have been eighteen cases of inhalation anthrax in the United States (4). Thirteen of these cases have been related to industrial exposures either directly or indirectly. A single case has occurred in a home weaver who was working with imported wool/hair yarn from Pakistan which was later shown to be positive for B. anthracis (5).

Inhalation anthrax is virtually impossible to diagnose on clinical grounds. Patients present with influenza-like symptoms which rapidly deteriorate into a septicemic picture with schock and death within a few days (6). The hallmark of the disease is the presence of large hilar lymph nodes on chest roentgenogram. The anthrax bacillus is susceptible to penicillin but therapy is usually begun so late as to be ineffective. Human cases have been well studied pathologically and have showed hemorrhagic mediastinitis and occasional cases of hemorrhagic meningitis. No ulcerative lesions have been found in the respiratory tract.

To better understand the relationship of the hair cloth manufacturing process to the aerosolization of anthrax spores, it is important to describe the early phase of the manufacture of hair cloth (7). After the bales are broken open, the goat hair is fed into a picking machine which helps break up the clumps of hair. The material then moves by conveyor apron to an enclosed duster where the hair is agitated by a high speed cylinder to loosen and shake dirt from the fibers. Waste material, dirt, dust and fibers which do not settle to the bottom of the duster are blown to a dust pit. This process is dusty and heavy aerosols are produced. After either wool or synthetic material has been combined with the cleaned goat hair the resultant blend is loaded into large carding machines which align the fibers and produce a loose rope of material. Carding is at times dusty but not as heavy as that in the picking and blending area. The material is then sent to the combing department where the short hairs are removed; after combing, the hair is further blended with wool and spun onto bobbins. The thread is then woven into hair cloth, which is finished by singeing, dying, washing and sanforizing. The risk of cutaneous or airborne anthrax is, to no great surprise, greatest in the early stages of the process.

The only epidemic of inhalation anthrax in the United States occurred in 1957 when five cases were reported over a ten week period in workers at a goat hair processing plant in Manchester, New Hampshire (6,8). No common predisposing illnesses were found among the patients. A temporal association was found between the onsets of the cases and the processing of a particular lot of Indian goat hair. It was presumed that the lot may have been unusually contaminated with anthrax spores since air sampling studies at several goat hair processing plants showed that workers routinely inhaled anthrax spores. Why some persons became ill while others, with presumably the same exposure, remained well, has never been adequately explained.

Pathophysiologic Studies

Anthrax infection of the lung has fascinated experimental microbiologists for many years. For several years there was controversy about the pathogenesis of inhalation anthrax and some felt that infection began with an ulcerative intrabronchial lesion, analogous to the sore of cutaneous anthrax. The question was decisively settled experimentally. Guinea pigs were inoculated intratracheally with anthrax spores, sacrificed over time, their lungs fixed and stained and examined microscopically. Spores were shown to be rapidly phagocytized by alveolar macrophages and transported to regional lymph nodes where germination of spores occurred. These vegetative forms overwhelmed local defenses and a bacillemia developed. These results showed that no true pneumonia developed but that the primary infection was an aggressive mediastinitis.

The above studies were useful in clarifying the fate of anthrax bacilli which reach the alveolar level by aspiration but were not airborne challenges. Druett and coworkers developed techniques whereby anthrax spores of varying sizes could be reliably prepared (9). They then exposed guinea pigs and monkeys to aerosol challenges of _B. anthracis_ ranging in size from 3.5 to 12 microns. Results of the guinea pig challenges are summarized in Table 1. As particle size increased, infectivity fell from 100 percent to 25 percent. Primate studies yielded similar results with smaller particles being fifteen times more infectious than larger particles. These data were entirely consistent with other information correlating site of deposition of aerosolized bacteria within the lung with aerodynamic diameter. Large particles as a rule are either filtered and removed in the nasopharynx or deposited on the bronchial mucociliary blanket, whereas particles in the range of 2 to 4 microns are capable of penetrating beyond ciliated epithelium to alveolar ducts and alveoli.

TABLE 1--Influence of size of inhaled _B. anthracis_ on mortality

Particle diameter (microns)	Number exposed	Number died	Percent mortality
3.6	40	40	100.0
8.4	40	11	27.5
11.6	40	10	25.0

Data taken from reference 9.

Albrink and Goodlow studied the effect of inoculum size by exposing primates to varying doses of aerosolized anthrax spores (10). Two chimpanzees exposed to a dose of 35,000 spores remained well despite sporadic positive blood cultures through the eleventh day after exposure. Two of four chimpanzees exposed to higher numbers of spores (40,000 to 100,000) became progressively bacteremic and died. At autopsy no pneumonia was found in either animal but rather an extensive hemorrhagic mediastinitis was found. In all such experiments higher inoculums were associated with higher mortality rates.

The most comprehensive and definitive study of the pathophysiology of industrial inhalation anthrax was a field experiment whereby monkeys were exposed to naturally occurring aerosols containing B. anthracis from a goat hair processing plant. These experiments were the research equivalents of the Manchester, New Hampshire epidemic of inhalation anthrax (11). The aerosol from the picking machine was sucked into a hood over the machine and carried through an exposure chamber to monkeys which were located outside the mill in a self-contained trailer. During the exposures air samples were constantly collected. One quarter of the 91 monkeys so exposed died of inhalation anthrax. Anthrax mortality rates were directly related to the numbers of spores inhaled. Monkeys inhaling about 17,000 viable spores had a fatality rate of 37 percent, while only 7 percent died after an exposure of 1,300 spores. The pathologic findings showing a bacillemia and hemorrhagic mediastinitis were identical to those noted in animals dying after exposure to pure aerosols of B. anthracis and humans dying of inhalation anthrax after industrial exposure.

SAMPLING METHODS

Air Sampling

The epidemic of inhalation anthrax in 1957 prompted the study of and quantitation of airborne anthrax particles in an industrial setting (12). Initial measurements of airborne microbial contamination was done by the U.S. Army Chemical Corps from Fort Dietrich. Cascade impactors (Anderson) and slit samplers were used in these studies as well as all glass impingers with British preimpingers. Suffice it to say that sampling in such mills was difficult because of the large amounts of dust and hair. Anthrax bacilli were often obscured by the presence of other closely related spore-forming organisms. Nonetheless, the principal aim of these studies was to quantitate aerosolized anthrax spores with particular attention to those less than 5 microns in size.

In general the cascade impactors using blood agar plates were run from 1 to 5 minutes in order that the plates were not overcrowded. All plates were incubated at 37C for 16 to 24 hours and potential anthrax colonies identified according to standard morphologic criteria. Subcultures were taken and final identification of anthrax bacilli was made according to standard microbiologic criteria. Under such circumstances anthrax made up less than one percent of the total viable organisms (7,12). Some investigators have tried to use a selective medium with propamidine and polymyxin B but have noted that anthrax colonies were always lower in number when the selective medium with used. When only bacteria less than five microns were counted, it was estimated that during an eight hour day a worker in the carding area would inhale between 140 to 690 anthrax spores less than five microns in size (12). These findings show quite clearly that workers in a goat hair processing plant frequently inhaled anthrax spores yet Woolsorters' disease was always a rare disease. Despite several attempts, no common predisposing factor among cases of Woolsorters' disease has been identified.

Environmental Samples

Environmental samples simply reflect the extent of surface contamination with *B. anthracis*. These samples are easy to perform and have been very useful in establishing the presence of anthrax spores in the environment. Moistened cotton swabs are swabbed over whatever surface needs to be sampled. These swabs are subcultured onto blood agar, incubated at 37C for 16 to 24 hours and examined for the presence of anthrax colonies. Results of environmental samples in a 1966 epidemiologic study of a case of inhalation anthrax in Manchester, New Hampshire are summarized in Table 2 (7). Anthrax bacilli were commonly recovered from swab samples taken from machines and the environment during the early phases of the manufacturing process. As the manufacturing process proceeded there were fewer isolations of anthrax bacilli, a finding consistent with the epidemiologic observation that workers in the early phases of the manufacturing process were at greater risk from cutaneous anthrax.

TABLE 2--Results of environmental samples
for the presence of *Bacillus anthracis*
Goat hair processing mill, Manchester, NH - 1966

Area	Swabs taken	Pos. *B. anthracis*	Percent pos.
Picking & blending	86	19	11.6
Carding	120	34	28.3
Combing	50	11	22.0
Weaving	50	1	2.0
Spinning & roving	100	3	3.0
Finishing	45	0	0.0
Total	451	68	15.1

CONTROL OF INHALATION ANTHRAX

Selective Decontamination and Anthrax Vaccine

Since it was impossible to guarantee bacteriologically clean raw material to industries manufacturing hair cloth, there were only two practical alternatives which could be done to protect workers; (1) decontaminating raw material after importation or (2) protecting workers with an anthrax vaccine. Both strategies were carried out successfully (13,14). The English built an anthrax decontamination station at Liverpool and all imported hair was decontaminated on arrival prior to its distribution to textile plants. No case of inhalation anthrax has been reported since this facility was built. The industrial anthrax problem in the United States was never a great one and the construction of a decontamination facility was not felt to be cost effective. Instead, an anthrax vaccine was developed and was shown to prevent cutaneous anthrax. All industrial workers at risk from anthrax-contaminated raw material were immunized. No fully immunized worker has developed inhalation anthrax, although the efficacy of the vaccine in specifically preventing inhalation anthrax remains to be proven.

OTHER GRAM POSITIVES

Several studies have demonstrated the presence of other Gram positives in the air, including streptococci, staphylococci, other bacillus and corynebacteria. Micrococci and staphylococci are commonly isolated. In the case of Staphylococcus aureus some persons have been shown to be shedders such that they release, presumably through skin shedding, staphylococci that may be isolated on settling plates, impaction devices and in liquid samples. Several epidemics of staphylococcal post-operative wound infections have been traced to such individuals in medical facilities. However, these organisms do not cause pneumonia after they are inhaled. The pathogenesis of staphylococcal and streptococcal pneumonia begins with oropharyngeal colonization, aspiration, local multiplication and the subsequent development of pneumonia. This process usually is superimposed on a preexisting respiratory viral infection like influenza. A great deal of clinical and experimental data support these concepts. Hence there is no need to perform sophisticated aerosol sampling studies in the setting of staphylococcal or streptococcal pneumonia except under the most unusual epidemiologic setting of multiple cases linked temporally and geographically.

REFERENCES

[1] Feeley, J.C. and Patton, C.M., "Bacillus" in Manual of Clinical Microbiology, ed 3. Washington, D.C., American Society for Microbiology, 1980, pp. 145-155.
[2] Koch, R., "Investigations of Bacteria: The Etiology of Anthrax, Based on the Ontogeny of the Anthrax Bacillus," Medical Classics, 1937-1938, Vol. 2, pp. 787-820.
[3] LaForce, F.M., "Woolsorters' Disease in England," Bulletin New York Academy of Medicine, 1978, Vol. 51, pp. 956-963.
[4] Brachman, P.S., "Inhalation Anthrax," Annuals New York Academy of Science, 1980, Vol. 353, pp. 83-93.
[5] Suffin, S.C., Carnes, W.H. and Kaufmann, A.F., "Inhalation Anthrax in a Home Craftsman," Human Pathology, 1978, Vol. 9, pp. 594-597.
[6] Plotkin, S.A., Brachman, P.S., Utell, M. et al, "An Epidemic of Inhalation Anthrax, the First in the Twentieth Century," American Journal of Medicine, 1960, Vol. 29, pp. 992-1001.
[7] LaForce, F.M., Bumford, F.H., Feeley, J.C. et al, "Epidemiologic Study of a Fatal Case of Inhalation Anthrax," Archives of Environmental Health, 1969, Vol. 18, pp. 798-805.
[8] Brachman, P.S., Plotkin, S.A., Bumford, F.H. et al, "An Epidemic of Inhalation Anthrax: The First in the Twentieth Century. II. Epidemiology," American Journal of Hygiene, 1960, Vol. 72, pp. 6-23.
[9] Druett, H.A., Henderson, D.W., Packman, L. et al, "Studies of Respiratory Infection. I. The Influence of Particle Size on Respiratory Infection with Anthrax Spores," Journal of Hygiene, 1953, Vol. 51, pp. 359-371.
[10] Albrink, W.S. and Goodlow, R.J., "Experimental Inhalation Anthrax in the Chimpanzee," American Journal of Pathology, 1959, Vol. 35, pp. 1055-1065.
[11] Brachman, P.S., Kaufmann, A.F. and Dalldorf, F.G., "Industrial Inhalation Anthrax," Bacteriology Review, 1966, Vol. 30,

pp. 646-657.
[12] Dahlgren, C.M., Buchanan, L.M., Decker, H.M. et al, "Bacillus anthracis Aerosols in Goat Hair Processing Mills," American Journal of Hygiene, 1960, Vol. 72, pp. 24-31.
[13] "Wool Disinfection and Anthrax," Lancet, 1922, Vol. 2, pp. 2:1295-1296.
[14] Brachman, P.S., Gold, H., Plotkin, S.A. et al, "Field Evaluation of a Human Anthrax Vaccine," American Journal of Public Health, 1962, Vol. 52, pp. 632-645.

DISCUSSION

Question: There are reports of two newly discovered species of corynebacteria causing pneumonia. Both appear to be water associated. Do you think that aspiration following drinking contaminated water is a more likely explanation than airborne transmission for these diseases?
Closure: Aspiration appears to be the far more common route of infection for most bacterial species. However, I am unaware of reports that have epidemiologically linked Corynebacterial infections to potable water.

Question: What is the best method to collect surviving Gram positive bacteria in water systems that have been chlorinated?
Closure: Chlorine levels decrease over time and no special microbiologic techniques are recommended.

Question: Is it possible for illnesses such as diphtheria to be transmitted by a building HVAC system?
Closure: The growth requirements of diphtheria are such that it is highly unlikely that they could be spread in a ventilating system.

Question: Recently there has been an increase in invasive streptococcal disease; could this be explained by these organisms surviving airborne transmission and reaching deeper parts of the respiratory tree?
Closure: There have been no reports of streptococcal pneumonia as part of the resurgence of disease due to Group A streptococci. Most of the illnesses have been pharyngitis with the later development of rheumatic fever.

Question: Some human-shed bacteria can survive as aerosols; is there any health significance to this phenomenon?
Closure: Carefully controlled studies have been unable to demonstrate any deleterious effect from the usual airborne micrococci or staphylococci; rather, their presence has served as a marker of crowding.

Question: What role might Gram positive bacteria play in subtle disease, since these bacteria produce many exotoxins known to have endotoxin-like activities?
Closure: As sensitising organisms airborne Gram positives might well be causative agents in cases of hypersensitivity pneumonitis.

Question: What is the best method of collection of human shed bacteria from the air? What effect, if any, does the humidity of the room air have on recovery?
Closure: Methods for sampling airborne microorganisms have been pub-

lished in *Applied Industrial Hygiene*, 1986, Vol. 4, pp. R19-R24. The author is not aware of experiments that have specifically studied the effect of humidity, although it has been generally held that microorganisms survive better in high humidity.

Mary A. Hood

GRAM NEGATIVE BACTERIA AS BIOAEROSOLS

REFERENCE: Hood, M.A., "Gram Negative Bacteria as Bioaerosols," Biological Contaminants in Indoor Environments, ASTM STP 1071, Philip R. Morey, James C. Feeley, Sr., and James A. Otten, Editors, American Society for Testing and Materials, Philadelphia, 1990.

ABSTRACT: Gram negative bacteria as possible etiological agents in respiratory diseases and syndromes associated with heating, ventilation and air conditioning (HVAC) systems and humidifying systems are reviewed. A case study of an outbreak of hypersensitivity pneumonitis possibly linked with a gram negative bacterium, *Cytophaga*, is presented, and sampling and culturing procedures for the enumeration, isolation and identification of gram negative bacteria are discussed.

KEYWORDS: bacteria, gram negative, respiratory diseases and syndromes, bioaerosols

Gram negative bacteria have been associated with two types of respiratory diseases and syndromes where exposure involves heating, ventilation and air conditioning (HVAC) systems and humidifying systems: 1) respiratory infections such as pneumonia where the bacterium enters the respiratory tract, colonizes by adhering and growing, and establishes a true infection and 2) respiratory symptoms which involve allergic type or immune reactions such as a) allergic rhinitis and asthma (although the allergens are mostly reported to be fungal), b) chronic bronchitis (which may be associated with endotoxins from gram negative bacteria), c) byssinosis or organic dust toxic syndrome (which is associated with exposure to a variety of fungi, bacteria, mycotoxins and endotoxins [1]), d) humidifier fever (HF) (which may be a type of byssinosis and is associated with protozoans, fungi and the endotoxin of gram negative bacteria occurring in water, originally humidifiers), and e) hypersensitivity pneumonitis (HP) (called extrinsic allergic alveolitis which is a T-lymphocyte dependent granulomatous inflamatory reaction and has been associated with fungi, actinomycetes, bacteria and protozoans)[2]. Organic dust toxic syndrome (and especially HF) share some of the same features of HP and

Professor, University of West Florida, Department of Biology
11000 University Parkway, Pensacola, FL 32514

there is considerable overlap among these diseases and syndromes. HF has generally been characterized by fevers, chills, muscle aches and malaise (for a review of the history of HF see references 3 and 4) while HP is a more serious lung disease characterized by acute recurrent pneumonia with fever, cough, chest tightness, shortness of breath, and fatigue. Diagnosis of HP is made by the physician on the basis of tests such as 1) restrictive pulmonary function tests, 2) exercise tolerance tests resulting in responses similar to patients with interstitial lung disease, 3) granulomas and interstitial fibrosis on biopsy, 4) with bronchial challenge of extracts of the organisms the reoccurrence of the symptoms and 5) response to corticosteroids [5,6]. The kinds of gram negative bacteria that have been implicated in such respiratory diseases and syndromes are presented in Table 1. Not included in the list are the bacteria that cause nosocomial (hospital acquired) respiratory infections or those associated with the compromised host. Genera such as *Proteus*, *Providencia*, other Enterobacteriaceae, *Moraxella*, *Alcaligenes*, and other Pseudomonas-related organisms have long been known to cause nosocomial pneumonia (infections).

TABLE 1 -- Gram Negative Bacteria Associated With Respiratory Diseases and Syndromes In Which Aerosolization Occurred By Heating, Ventilation and Air Conditioning Systems or Humidifying Systems.

Organism	Symptoms/Disease	Reference
Enteric Fermenters		
Serratia marcescens	cough, fever, malaise	[7]
Klebsiella pneumoniae	classical bacterial pneumonia (infection)	
Klebsiella sp.	HF[a]	[8]
Enterobacter aerogenes	HF	[9]
Enterobacter sp.	HF	[8]
Salmonella typhimurium	pneumonia (infection)	[10]
Non-enteric Fermenters		
Aeromonas sp.	HF (unconfirmed)	[11]
Pseudomonas related organisms		
Pseudomonas aeruginosa	pneumonia (infection)	[12,13]
Pseudomonas sp.	HF	[8]
Flavobacterium sp.	HF	[14]
F. meningosepticum	pneumonia (infection)	[15]
Cytophaga allerginae	HD[b]	[16]
Achromobacter sp.	HF (unconfirmed)	[11]
Acinetobacter	infection	[17]
Others		
Legionella	pneumonia (infection)	(see article by Dennis, this volume)
Brucella	laboratory acquired infection	[18]

[a]HF = Humidifier fever associated with the endotoxin of the bacterium
[b]HP = Hypersensitivity pneumonitis

Many gram negative bacteria survive and grow well in aquatic environments which makes them ideal inhabitants of HVAC systems or humidifiers where water accumulates and/or is aerosolized. Gram negative bacteria possess several characteristics which may make them ideal to act as allergens or infectious agents. Table 2 presents these characteristics.

TABLE 2 -- Characteristics of Gram Negative Bacteria that may Act to Enchance Properties as Potential Problematic Bioaerosols

1. Many gram negative bacteria range in size from 1 to 5 μ in length and in nutrient dilute water may exhibit as much as 90% reduction in volume. This allows penetration into the lung.

2. All species possess the **(LPS) Lipopolysaccharide layer** composed of side polysaccharides, core polysaccharides and Lipid A [endotoxin].

3. Many species produce **surface proteins and slime layers or glycocalyx** (which helps in adherance; protection from drying, biocides, etc.) [other possible bioactive allergens].

Since many gram negative bacteria are so small and even more reduced in volume when in low nutrient waters, they may be well suited for deep penetration into the lung upon aerosolization. Secondly, all gram negative bacteria possess an endotoxin which is part of the LPS of their cell wall. The role of bacterial endotoxin is emerging as an important factor in respiratory diseases and syndromes [1]. In addition, surface proteins (in a variety of structures such as flagella, pilli, and cell wall) and slime layers or the glycocalyx which is composed of carbohydrates and/or proteins may act as allergens. While little is known concerning the bioactivity of these structures in respiratory diseases, they are well recognized antigens and may be important components in allergic responses. Certainly this is an area that needs study.

CASE STUDY

In the mid-1970's, an outbreak of hypersensitivity pneumonitis occurred among plant workers in a nylon textile plant [19]. The cases occurred only where the air was cooled and humidified by air-washing units which consisted of pumps that sprayed water into the air. The mist collected on demister vanes, ran down into a collecting sump, was returned to the reservoir and was then recycled to the spray. A slime developed in the reservoirs, sumps, and demister vanes. Using a solid-phase radioimmunoassay, the HP antigen was detected in high levels in this slime [20].

Using standard microbiological procedures, bulk samples of the slime were collected during the 60 day operational cycle, diluted and homogenized in phosphate buffered water plus glass beads, and innoculated onto the surface of ten different media [(Difco): plate count agar, nutrient agar, 1/2 strength nutrient agar, 1/10 N.A., blood agar (Biocon), MacConkey's, EMB, Endo, Oatmeal and Cytophaga Agar [21]. Incubation was carried out at 8, 20, and 35°C at 21, 14, and 4 days. The different colonies were restreaked and each isolate was subjected to a series of biochemical tests [22,23].

Twenty eight of the predominant isolates were examined for their cross-reactivity with the HP antigen and one isolate showed a highly positive reaction [24]. However, the precipitin data collectively suggested that this organism might not be the major antigen [20]. The isolate was identified as a species of *Cytophaga*. Although phage typing by Dr. Jack Pate (personal communication) revealed a phage profile similar to *Cytophaga johnsoniae*, our biochemical data, fatty acid and carbohydrate profiles and G + C ratios suggested a species different than *C. johnsoniae* or *C. aquatilis*. On the basis of our data, we proposed the name *Cytophaga allerginae* [22].

Human challenge tests using *Cytophaga allerginae* reproduced the symptoms of HP in a patient (Flaherty, personal communication) suggesting that the bacterium may play a role in the HP outbreak. However, it has been suggested that challenge with (non-specific) lipopolysaccharide would likely provoke a similar response. Clearly, the problems involved in "proving" the etiological agent in respiratory diseases such as HP (or HF or others) are complex. Further investigations are certainly warranted with regard to this organism's role in HP as well as the role of other gram negative bacteria in HP and HF.

SAMPLING AND CULTURING METHODS

In addition to humidifiers, cooling units and other water-containing systems, ruptured sanitary drains and sewage lines may also be the source of gram negative bacteria. If such material gets into the HVAC, the bacteria can become aerosolized, and exposure may lead to air quality problems.

Two reports [5,25] developed by the Bioaerosol Committee of the American Conference of Governmental Industrial Hygienists (ACGIH) provide specific guidelines for sampling bioaerosols. After a walk through inspection to identify possible reservoirs and amplifications of microbial aerosols, air sampling is recommended if all other factors are eliminated (such as chemicals, etc.).

There are a number of air samplers available [5,26] including slit samplers [27], sieve type impactors [28], filter cassettes, glass impingers [29], and centrifugal samplers [27]. The Andersen viable impactor has generally been accepted as the sampler for use especially when bioaerosols are low [25]. The media recommended for culturing bacteria, in general, are typticase soy agar [25] and brain heart infusion agar [30]. However, the organisms may not be recovered on full strength media and require diluted media for growth. For specific bacteria such as the gram negative bacteria, media such as MacConkey, EMB, Endo or other selective media might be useful. However, there are no studies which evaluate these media with respect to recoverability of gram negative bacteria in the aerosol state. Likewise, individual organisms such as Pseudomonas sp. can be cultured by using numerous selective media such as King's Medium [31], modifications of this medium, Bacto Pseudomonas Agar F and P (Difco), Bacto Pseudomonas Isolation Agar (Difco), and cycloheximide agar [32] as well as various modifications [21]. Cytophaga sp. can also be selected using medium with added compounds toxic for other organisms. Molds, for example, can be eliminated using cycloheximide; actinomycetes, by adding neomycin and polymyxin B. (A list of these media are presented and discussed in Bergey's Manual Vol. 3, pp 2024-2029. [21]). Again, no studies have been conducted to evaluate any of these media in terms of recovering aerosolized organisms. Perhaps the best recommendation when a specific gram negative bacterium is suspected is to employ the media used to recover that organism from water or if there are no procedures for environmental samples to use the protocol for recovery of that organism from clinical specimens.

Incubation of media is recommended at 35°C, although if the suspect water source is a cool or ambient temperature the microorganisms may not grow at 35°C and a room temperature or lower incubation may be required. CFU (colony forming units) are determined and expressed as CFU/m^3 (CFU m^{-3}). High levels of gram negative bacteria (5.0 X 10^2CFU m^{-3}) may be a presumptive indication of indoor air quality problems and remedial actions would be prudent [25]. It should be pointed out that these values are somewhat arbitrary since systematic studies correlating numbers of bacteria with respiratory outbreaks have not been conducted.

Identification of the dominant organisms is also recommended, i.e., those in concentrations higher than 75 CFU m^{-3} [5]. Gram negative bacteria belong to three taxonomic groups: 1) the aerobic rods and coccobaccilli which exhibit strictly respiratory type metabolism (oxidative), 2) the facultatively anaerobic rods and coccobaccilli which exhibit aerobic and anaerobic metabolism (oxidative-fermentative), and 3) bacteria with unusual properties. The aerobic and facultatively anaerobic rods are described in Volume 1 of Bergey's Manual of Systematic Bacteriology [21] while those with unusual properties such as the gliding bacteria (Cytophaga, Flexibacter, Beggiatoa) the budding and/or appendaged bacteria (Caulobacter), and the sheathed bacteria (Sphaerotilus) are described in Volume 3. Taxonomically, gram negative bacteria can be identified using a number of schemes which make use of biochemical/morphological tests. After the gram stain, oxidation-fermentation reactions in glucose and the oxidase test will place the organism into one of four groups: 1) glucose fermenters-oxidase positive, 2) glucose fermenters-oxidase negative, 3) non-fermenters-oxidase positive, and 4) non-fermenters-oxidase negative. Group 2 comprise the 20 genera of the family Enterobacteriaceae which include the genera Escherichia, Enterobacter, Shigella, Salmonella, etc. These organisms have been well studied clinically and can be differentiated to species level on the basis of a number of biochemical tests [33]. Numerous rapid identification systems have been developed for the clinical isolates of this group. Group 1 includes the genera Plesiomonas, Aeromonas, Chromobacterium, Actinobacillus, and others. Again, biochemical and morphological tests must be used to differentiate the organisms to genus level. Group 3 are the Pseudomonas related organisms and include the genera Pseudomonas, Flavobacterium, Alcaligenes, etc., while group 4 include

Acinetobacter and Acetobacter. Again, an examination of Bergey's [21] can provide the microbiologist with the appropriate tests to use for identification to genus level. It should also be mentioned that a number of gram negative bacteria will not grow on nutrient media such as typticase soy or brain heart infusion. These fastidious organisms such as Haemophilus, Brucella, Francisella as well as Legionella require special media and procedures for their recovery. Finally, remedial action can be summarized in the following table.

TABLE 3 -- Recommendations for Remedial Actions Against the Amplification of Gram Negative Bacteria [5]

1. Remove and prevent the accumulation of (especially stagnant) water and slimes from mechanical systems. When the system is off, disinfect with a chlorine solution. Steam cleaning or abrasive mechanical cleaning may be necessary.
2. Eliminate water spray systems in air handling units if possible.
3. Substitute steam for cold water humidifiers if possible.
4. Eliminate the use of cold mist vaporizers if possible.
5. Keep indoor relative humidity below 70%. For locations where cold surfaces are in contact with warm air, keep indoor relative humidity below 50% if possible.
6. Replace HVAC system filters and old components at scheduled intervals.
7. If water must be used, it must be kept clean and low in nutrients. Biocides might be needed [34]. Since gram negative bacteria, especially Pseudomonas species, are known to be more resistant to chemical antimicrobial agents than other microorganisms, special care must be taken to select the appropriate biocide.

In summary, gram negative bacteria are emerging as possible etiological agents of respiratory diseases and syndromes in HVAC and humidifying systems, but there is clearly a lack of information regarding the exact role of these bacteria in such diseases. Since they comprise the natural biota of water bodies and their concentrations can be amplified with manipulations such as humidifiers and various cooling systems, it is important to recognize them and to develop techniques for enumerating, culturing and identifying them. Since there are few studies which address the issue of recovering gram negative bacteria from aerosols this is an area which needs rigorous methods investigations. It is also likely that as more and more cases of respiratory diseases and syndromes are examined, gram negative bacteria may be implicated as we begin to look for them. Thus, more studies will be needed to develop better methods as well for their control.

REFERENCES

[1] Rylander, R. and Burrel, R. "Endotoxins in Inhalation Research. Summary of Conclusions of a Workshop Held at Clearwater Florida." Annual Occupational Hygiene. Vol. 32, No. 4, 1988, pp 553-556.
[2] Lacey, J. and Crook, B. "Review: Fungal and actinomycetes spores as pollutants of the workplace and occupational allergens," Annual Occupational Hygiene, Vol. 32, No. 4, 1988, pp 513-533.
[3] Ager, B.P., and Tichner, J.A., "The Control of Microbiological Hazards Associated with Air Conditioning and Ventilation Systems," Annual Occupational Hygiene, Vol. 27, No. 4, 1983, pp 341-358.
[4] Edwards, J.H., "Humidifier Fever," Royal Society of Health Journal, Vol. 102, No. 1, 1982, pp 7-9.
[5] Burge, H.A., Chatigny, M., Feeley, J., Kreiss, K., Morey, P., Otten, J., and Peterson, K., "Bioaerosols: Guidelines for Assessment and Sampling of Saprophytic Bioaerosols in the Indoor Environment," Applied Industrial Hygiene, Vol. 2, No. 5, 1987, pp R-10-R-16.
[6] Solomon, W.R., and Burge, H.A., "Allergens and Pathogens," Indoor Air Quality, P.S. Walsh, C.S. Dudney, E.D. Copenhauer, Eds., CRC Press, Boca Raton, FL, 1984, pp 173-191.
[7] Cole, L.A., "Clouds of Secrecy: The Army's Germ Warfare Tests Over Populated Areas," Rowman and Littlefield Publishers, New Jersey, 1988.
[8] Helander, I., Salkinoga-Salonen, M., and Rylander, R., "Chemical Structure and Inhalation Toxicity of Lipopolysaccharides from Bacteria on Cotton," Infection and Immunity, Vol. 29, 1980, pp 859-862.
[9] Rylander, R., and Snella, M., "Acute Inhalation Toxicity of Cotton Plant Dusts," British Journal of Industrial Medicine, Vol. 33, 1976, pp 175-180.
[10] Leedon, J.M., and Loosli, C.G., "Airborne Pathogens in the Indoor Environment With Special Reference to Nosocomial (Hospital) Infections," Aerobiology: The Ecological Systems Approach, Edmonds, R., Ed., Dowden, Hutchinson and Ross, Stroudsburg, PA, 1979.
[11] Hollick, G.E., Larsh, H.W., Hubbard, J.C., and Hall, N.K., "Aerobiology of Industrial Plant Air Systems: Fungi and Related Organisms," Medical Mycology Zbl. Bakt. Suppl. 8, Fischer Verlag, New York, 1980, pp 89-95.
[12] Doggett, R.G., Ed., "Pseudomonas Aeruginosa: Clinical Manifestations of Infection and Current Therapy," Academic Press, New York, 1979.
[13] Grieble, H.G., Colton, F.R., Bird, T.J., Toigo, A., and Griffith, L.G., "Fine-Particle Humidifiers, Source of Pseudomonas Aeruginosa Infection in a Respiratory-Disease Unit," New England Journal of Medicine, Vol. 282, 1970, pp 531-535.
[14] Rylander, R., Haglind, P., Lundholm, M., Mattsby, I., and Stenquist, K., "Humidifier Fever and Endotoxin Exposure," Clinical Allergy, Vol. 8, 1978, pp 511-516.
[15] Von Graevenitz, A., "Clinical Significance and Antimicrobial Susceptibility of Flavobacterium," In Reichenback and Weeks, Eds., The Flavobacterium-Cytophaga Group. Proceedings of the International Symposium on Yellow-Pigmented Gram-Negative Bacteria of the Flavobacterium-Cytophaga Group, Verlag Chemie, Weinheim, 1981, pp 153-164.

[16] Liebert, C.A., Hood, M.A., Deck, F.H., Bishop, K., and Flaherty, D.K., "Isolation and Characterization of a New Cytophaga species Implicated in a Work-Related Lung Disease," Applied and Environmental Microbiology, Vol. 48, No. 5, 1984, pp 936-943.
[17] Smith, P.W., "Room Humidifiers as a Source of Acinetobacter Infections," Journal of American Medical Association, Vol. 237, 1977, p 795.
[18] Buchanan, T.M., Hendricks, S.L., Patton, C.M., and Feldman, R.A., "Brucellosis in the U.S. 1960-1972: an Abattoir-Associated Disease. III. Epidemiology and Evidence for Acquired Immunity," Medicine, Vol. 53, 1974, pp 427-439.
[19] Flaherty, D.K., Deck, F.H., Cooper, J., Bishop, K., Winzenburger, P.A., Smith, L.R., Bynum, L.M., and Witmer, W.B., "Bacterial Endotoxin Isolated from a Water Spray Air Humidification System as a Putative Agent of Occupation-Related Lung Disease," Infection and Immunity, Vol. 43, 1984, pp 206-212.
[20] Reed, C.E., Swanson, M.C., Lopez, M., Ford, A.M., Mayor, J., Witmer, W.B., and Valdes, T.B., "Measurements of IgG Antibody and Airborne Antigen to Control an Industrial Outbreak of Hypersensitivity Pneumonitis," Journal of Occupational Medicine, Vol. 25, No. 3, 1983, pp 207-210.
[21] Bergey's Manual of Systematic Bacteriology, Vols. 1 and 3, Williams and Wilkins, Baltimore, MD, 1984, 1989.
[22] Liebert, C.A., Hood, M.A., Winter, P.A., and Singleton, F.L., "Observations on Biofilm Formation in Industrial Air-Cooling Units," Developments in Industrial Microbiology, Vol. 24, 1983, pp 509-517.
[23] Liebert, C.A., and Hood, M.A., "Characterization of Bacterial Populations In An Industrial Cooling System," Developments In Industrial Microbiology, Vol. 26, 1985, pp 649-660.
[24] Flaherty, D.K., Deck, F.H., Hood, M.A., Liebert, C.A., Singleton, F.L., Winzenburger, P.A., Bishop, K., Smith, L.R., Bynum, L.M., and Witner, W.B., "A Cytophaga Species Endotoxin as a Putative Agent of Occupation-Related Lung Disease," Infection and Immunity, Vol. 43, 1984, pp 213-216.
[25] Morey, P., Otten, J., Burge, H., Chatigny, M., Feeley, J., Laforce, F.M., and Peterson, K., "Bioaerosols Airborne Viable Microorganisms In Office Environments: Protocol and Analytical Procedures," Applied Industrial Hygiene, Vol. 4, No. 1, 1986, pp R-19-R-23.
[26] Groschel, D.H., "Air Sampling in Hospitals," Annual New York Academy of Science, Vol. 353, 1980, pp 230-239.
[27] Casewell, M.W., Fermie, P.G., Thomas, C., Simmons, N.A., "Bacterial Air Counts Obtained With a Centrifugal (RCS) Sampler and a Slit Sampler - the Influence of Aerosols," Journal Hospital Infections, Vol. 5, 1984, pp 76-82.
[28] Jones, W., Morring, K., Morey, P., and Sorenson, W., "Evaluation of the Andersen Viable Impactor for Single Stage Sampling," American Industrial Hygiene Association Journal, Vol. 46, No. 5, 1985, pp 294-298.
[29] Tyler, M.E., and Shipe, E.L., "Bacterial Aerosol Samplers I. Development and Evaluation of the All-Glass Impinger," Applied Microbiology, Vol. 7, 1959, pp 337-354.

[30] McGowan, J., Jr., "Role of the Microbiology Laboratory in Prevention and Control of Nosocomial Infections," Manual of Clinical Microbiology, 4th Ed., Lennette, Balows, Hausler, Shadomy, Eds., American Society for Microbiology, Washington, DC, 1985, pp 110-122.
[31] King, E.O., Ward, W.K., and Raney, D.E., "Two Simple Media for the Demonstration of Pyocyanin and Fluorescein," _Journal Laboratory Clinical Medicine_, Vol. 44, 1954, pp 301-307.
[32] Sands, D.C., and Rovira, A.D., "Isolation of Fluorescent Pseudomonds with a Selective Medium," _Applied Microbiology_, Vol. 20, 1970, pp 513-514.
[33] Lennette, E.H., Balows, A., Hausler, W.J., Jr., and Shadomy, H.J., "Manual of Clinical Microbiology," 4th Ed., American Society for Microbiology, Washington, DC, 1985.
[34] Sharpell, F., "Industrial Uses of Biocides in Processes and Products," _Developments In Industrial Microbiology_, Vol. 21, 1980, pp 133-140.

DISCUSSION
What is the best media and incubation procedure for cultivation of _Pseudomonas aeruginosa_? What is the best media and incubation procedure for collection of gram negative bacteria in general?
CLOSURE
There are numerous procedures used to enumerate _Pseudomonas aeruginosa_. One method used to determine the number of _P. aeruginosa_ as an indicator of the quality of recreational waters makes use of M-PA agar and is described in "Standard Methods for the Examination of Water and Wastewater (APHA, 1980). Other media such as King's Medium B and Medium A are used (see Bergey Manual of Determinative Bacteriology vol. 1) for environmental samples, while clinical laboratories employ blood agar and look for distinct colony types. Until more studies are conducted, it is difficult to recommend a best media or incubation procedure for gram negative bacteria but several might be in order: 1) MacConkey's, EMB and Endo media are probably the most widely used selective media for growing gram negative bacteria. 2) If water temperature are cold, it might be appropriate to incubate media at both 35 and 20°C.
DISCUSSION
Describe culture and incubation procedures for _Cytophaga_ species. Room temperature as well as psychrophiles. Do you think that _Cytophaga_ could be collected by an aerosol sampler?
CLOSURE
There is no one specific selective procedure for culturing all _Cytophaga_ species. They are too heterogenous. However, Bergey's Manual (vol. 3) gives a summary of procedures and references for culturing these organisms. It would probably be safe to say that a proteose peptone medium could be used to which compounds toxic to other organisms were added. _Cytophaga_ species are relatively resistant to polymyxins, the penicillins, the aminoglycosides and chloramphenicol. _Cytophaga_ species are typically mesophiles (with some psychrotrophiles) so could be grown at both ranges of 35°C to room temperature and 15°C to just above freezing. While _Cytophaga_ species are generally

considered aquatic, they could very well be aerosolized and survive. How long they survive as aerosols is unknown but they would not survive as well as gram positive organisms or fungi.

DISCUSSION

You mentioned the use of challenge tests in HP after isolation of the offending agent. How often is it justified to perform a life threatening test such as this to prove which organism was causal when by temporal sequence and high level of contamination one can make diagnosis in most cases?

CLOSURE

Good point and question. It is not justifiable to use human challenge tests for quality control problems. The idea is to keep humidifiers, air cooling units and HVAC systems free of contamination, i.e., high levels of gram negative bacteria. Using standard enumeration methods, one would develop monitoring protocol to insure adequate quality control. There would be no need to prove which specific organism was causing allergic reactions, HP, etc. The only time challenge tests should be used is for epidemiological research purposes and then there are ways around using humans.

DISCUSSION

On the 60 day humidifier cleaning schedule, an ecological succession was observed. Was this in the presence of chlorinated spray? If so do you think chlorine tolerant species were selected for in the succession?

CLOSURE

At the time of our study, the water used to humidify the yarn plant was being treated with a biocide. As I recall, it was an organic chlorine biocide. But it was applied in batch so I suspect that the succession observed on steel plates was not related to chlorine treatment but rather due to natural colonization processes. In the water itself, whose source was a local river, certainly changes occurred and were probably reflected in the microbial population shifts.

DISCUSSION

What is the best sampling method for collecting viable gram negative bacteria in drift or aerosol from an electric power plant that uses treated sewage effluent? What sampler? What media? Has anyone done that and what is interpretation of data?

CLOSURE

The Anderson sampler, N-6, is recommended for aerosol sampling although there are others available (see chapter for references). With regard to the media, two types might be employed: 1) a selective medium (EMB, MacConkey's, Endo) and 2) a plate count media (TSA, BHI, peptone based media). I am not aware of any studies that have been done examining electric power plant and sewage effluent.

DISCUSSION

What is the significance, if any, of finding 10^6 *Flavobacterium* species per gram of dried debris in an air handling unit drain pan in an office? In a hospital?

CLOSURE

Unlike clinical samples where the microbiologist looks for specific microorganisms which are known to cause infection, or even with samples where *Legionella pneumophila* is suspected, the relationship between gram negative bacteria as a group or specific gram negative bacteria in aerosols or water used for aerosols and disease is not well established. This sort of microbiological data (10^6 cells of *Flavobacterium* cells/gm) is significant only when combined with other information. If, for example, the area was the site of reported allergic reactions in office workers/hospital workers or patients (while other areas not reporting problems were free of Flavobacterium) there might be some reason to suspect that perhaps these Flavobacterium might be involved. However, considerably more information would have to be gathered before such a correlation could be made. By itself, the fact that 10^6 Flavobacterium cells/gm are present in dry material in a drain pan probably has little meaning. However, it would be prudent to keep these drain pans clean and to insure that the water draining into these pans was clean.

DISCUSSION

Yersinia enterocolitica is a gram negative bacterium that is present in many water systems in the northeast. Assuming that it enters a building HVAC system or potable water services, and you are asked to sample for it in both air and water, how would you do so?

CLOSURE

There are many procedures for culturing Y. enterocolotica from food samples. Selective media such as CAL medium, RS and some common Enterobacteriaceae media (such as MacConkey agar, SS, etc.) are used at cool incubation temperatures of approximately 28-29°C for 48 hours followed by room temperature incubation for 48-72 hours. Also several cold enrichments have have been published (see Bergey's Manual, 1984, vol. 1). Confirmation is then done using biochemical tests (we have used the rapid method, API-21E test strip). I imagine these procedures could be adapted for air and water samples. For example, water could be directly plated onto the selective media, or diluted and plated. If the water had to be filtered, levels would probably be so low as to suggest an insignificant role for the organism. Likewise, air samples could be taken and plated onto the media. Confirmation of the organisms based on selecting colony type and performing biochemical tests would then have to be performed.

Joseph O. Falkinham, III, Karen L. George, Mary A. Ford and Bruce C. Parker

COLLECTION AND CHARACTERISTICS OF MYCOBACTERIA IN AEROSOLS

REFERENCE: Falkinham, J. O., III, George, K. L., Ford, M. A., and Parker, B. C., "Collection and Characteristics of Mycobacteria in Aerosols," Biological Contaminants In Indoor Environments, ASTM STP 1071, P. R. Morey, J. C. Feeley, Sr., and J. A. Otten, Eds., American Society for Testing and Materials, Philadelphia, 1990.

ABSTRACT: Airborne mycobacteria in natural aerosols in Richmond, Virginia were isolated and enumerated using an Andersen Sampler and suitable bacteriologic media. *Mycobacterium avium*, *M. intracellulare* and *M. scrofulaceum* (the MAIS group) were recovered in higher numbers from aerosols than other mycobacteria, though there was wide fluctuation in the numbers of each, depending upon the collection site. Mycobacteria were isolated on particles of various sizes, some capable of reaching the alveoli of the human lung. *Mycobacterium avium*, *M. intracellulare* and *M. scrofulaceum* were recovered from waters upstream of Richmond, Virginia, whereas highest numbers of *M. terrae* and *M. gordonae* and rapidly growing mycobacteria were recovered from waters collected downstream of Richmond's municipal sewage treatment plant. Based upon the numbers of *M. avium*, *M. intracellulare* and *M. scrofulaceum* in aerosols and in the James River, the major pathway of human exposure would be through inhalation. The numbers of mycobacteria in aerosols did not correlate with the numbers in waters.

KEYWORDS: aerosols, mycobacteria, particle size, water

The epidemiology and aerosol transmission of the obligate human pathogen, *Mycobacterium tuberculosis* has been well documented [1]. By contrast, the epidemiology and route of human infection of members of the *M. avium*, *M. intracellulare* and *M. scrofulaceum* group (the MAIS group) is less well understood. Members of the MAIS group are closely related, opportunistic human pathogens whose source of human infection is the environment [2-5]. Members of this group of

Joseph O. Falkinham, III, Karen L. George, Mary A. Ford and Bruce C. Parker, Department of Biology, Virginia Polytechnic Institute and State University, Blacksburg, VA 24061.

bacteria cause pulmonary infections [2], have been recovered from natural aerosols collected in Richmond, Virginia on the shore of the James River [4] and are enriched in droplets aerosolized from suspensions [6]. Naturally aerosolized MAIS organisms appear to be the source of human MAIS infections because of the identity of isolates recovered from patients and aerosols [7,8]. Few soil, dust and water isolates resembled clinical isolates [8]. Within the past 5 years it has become evident that a significant proportion of patients with acquired immune deficiency syndrome (AIDS) have life-threatening MAIS infections [9]. Because MAIS organisms have been isolated and are concentrated in hospital hot water systems, it has been suggested that such hospital hot water systems could serve as the source of MAIS infection in AIDS and other immunocompromised patients [10]. As a first step of a study of the generation of indoor MAIS aerosols in hospitals, we have studied the annual pattern of mycobacteria in natural, outdoor aerosols to serve as a control. Our approach involved isolation of mycobacteria from James River aerosols and water samples collected at two week intervals in Richmond, Virginia during the period July 1979 through July 1980 using the Andersen 6-stage cascade sampler [11]. The objective was to measure the number and seasonal fluctuation of MAIS organisms in aerosols and determine whether correlations existed between numbers of mycobacteria in aerosols and in water samples.

EXPERIMENTAL METHOD

Collection Sites

Samples were collected approximately every two weeks from 12 July 1979 through 17 July 1980. Water and aerosol samples were collected at three sites along the shore of the James River in Richmond, Virginia [4]: (site 1) downstream of Richmond and below the outflow of the Richmond Municipal Sewage Treatment Facility, (site 2) in the center of the city at the Boulevard Bridge, and (site 3) upstream of the city at the Hugenot Bridge. Additional aerosol samples were collected monthly at sites 0.35 mi north (Byrd Park; Byrd) or south (Westover Hills Elementary School; West) of the James River. At each site, duplicate (and simultaneous) samples of both water and aerosols were collected. Also, duplicate aerosol samples were collected from the south (S) and north (N) shores of the Boulevard Bridge to judge the influence of wind direction on the numbers of mycobacteria in aerosols.

Aerosol Collection and Isolation of Mycobacteria

Aerosol samples were collected simultaneously using two six-stage Andersen cascade samplers [11] which received air at 0.028 m^3/min for 60 min (1.7 m^3). Particles were collected on the surface of Middlebrook 7H10 agar medium (BBL Microbiology Systems, Cockeysville, MD) containing 0.5% (vol/vol) glycerol. To prevent fungal overgrowth and enhance mycobacterial growth, the medium was prepared with 0.0025% (wt/vol) malachite green and 10% oleic acid-albumin-dextrose-catalase enrichment (BBL Microbiology Systems, Cockeysville, MD), respectively. Plates were used either 2 or 3 days after preparation. Use of freshly prepared medium (i.e. 1 day) yielded confluent growth of contaminants on the wet medium, while use of older (i.e. drier) medium resulted in an absence of mycobacterial and other bacterial colonies. Following collection, any liquid on the plates was spread vigorously to dryness with a sterile glass rod to disrupt clumps [6], the plates were placed in plastic sleeves and returned to the laboratory for incubation at 37°C for up to 8 weeks.

Ejected Droplet Collection, Droplet Diameter Measurement and Isolation of Mycobacteria

Droplets, ejected from the James River water surface were collected and mycobacteria were cultured as described by Wendt [4]. Any liquid (i.e. ejected droplets) on the exposed medium surface was spread vigorously to dryness. The diameter of ejected droplets and calculation of mycobacterial concentrations in ejected droplets was performed as described by Parker, et al. [6].

Water Collection and Isolation of Mycobacteria

Duplicate subsurface water samples of 1 liter were collected in sterile glass bottles during aerosol sampling. Mycobacteria in water samples were isolated and enumerated as described by Falkinham, et al. [3].

Identification of Mycobacteria

Colonies of acid-fast microorganisms appearing on any of the Middlebrook 7H10 agar media following 4-8 weeks incubation at 37°C were identified by standard mycobacterial methods [12]. In addition to the isolation, identification and enumeration of members of the MAIS group, the numbers of *M. terrae* and *M. gordonae* (NON-MAIS) and rapidly growing mycobacteria (RAPID) were also determined [3]. No other species of mycobacteria were recovered from aerosols or waters.

RESULTS

Mycobacteria in aerosols

MAIS numbers were higher in aerosols (23 ± 53 per 1.7 m^3 aerosol), compared to *M. terrae*/*M. gordonae* (NON-MAIS; 4 ± 11) and rapidly growing mycobacteria (RAPID; 7 ± 25). These data reflect means and standard deviations of 26 sampling dates. There was a wide variation in numbers of MAIS (Figure 1), *M. terrae*/*M. gordonae* and rapidly growing mycobacteria (data not shown) in aerosols collected at different times of the year. Duplicate samples collected at the same site and time yielded similar values. The only instance where the numbers collected at the north and south ends of the Boulevard Bridge did not agree was on August 16, 1980. The south (downwind) collection yielded 121 rapid growing mycobacteria/1.7 m^3, while the north (upwind) collection yielded only 5 rapid growing mycobacteria/1.7 m^3.

The variation in MAIS numbers was due to high numbers recovered from aerosols collected at either Byrd Park or Westover Hills Elementary School; of the 26 aerosol sample dates, 4 had greater than 10 MAIS/1.7 m^3 and 3 of those were collected at either the Park or School. Only on June 19, 1980, were high numbers of MAIS in aerosols collected at both the north and south shore of the James River at the Boulevard Bridge. Few mycobacteria were recovered from aerosols collected at the Sewage Treatment Facility.

When MAIS organisms were recovered from aerosols, *M. terrae*/*M. gordonae* and rapidly growing mycobacteria were recovered as well. No seasonal pattern in the number of MAIS organisms in either water or aerosol samples (Figure 1) was observed.

Particle size associated with mycobacteria in aerosols

In addition to its utility in the isolation and enumeration of viable microorganisms in aerosols, the Andersen 6-stage cascade sampler separates aerosol particles (e.g. water droplets and dust and soil particles) on the basis of size [11]. In Table 1 is listed the size of MAIS-associated particles recovered on August 16, 1979, March 27, 1980, May 22, 1980 and June 19, 1980. Only on June 19 were the majority of isolates recovered from the north and south banks of the James River at the Boulevard Bridge.

TABLE 1 - Size distribution of MAIS-associated particles

Date	Site	MAIS-Recovered from Particles of Diameter[a]					
		>8μ	5-10μ	3-6μ	2-4μ	1-2μ	<1μ
8/16/79	Bridge S	8	2	2	0	6	0
	Bridge N	5	1	1	1	0	0
	Byrd	19	35	5	1	5	6
3/27/80	Bridge S	1	0	1	0	0	0
	Bridge N	1	1	0	0	0	0
	West	58	24	5	0	5	1
5/22/80	Bridge S	1	0	0	0	0	0
	Bridge N	0	0	1	0	0	0
	West	0	57	15	6	2	1
6/19/80	Bridge S	95	9	3	0	0	1
	Bridge N	90	13	2	1	0	0
	Byrd	16	5	3	0	1	0

[a] Data expressed as MAIS colonies per 1.7 m^3 aerosol for particles of different diameter (expressed in microns) [11])

FIGURE 1. Recovery of MAIS organisms from water and aerosols

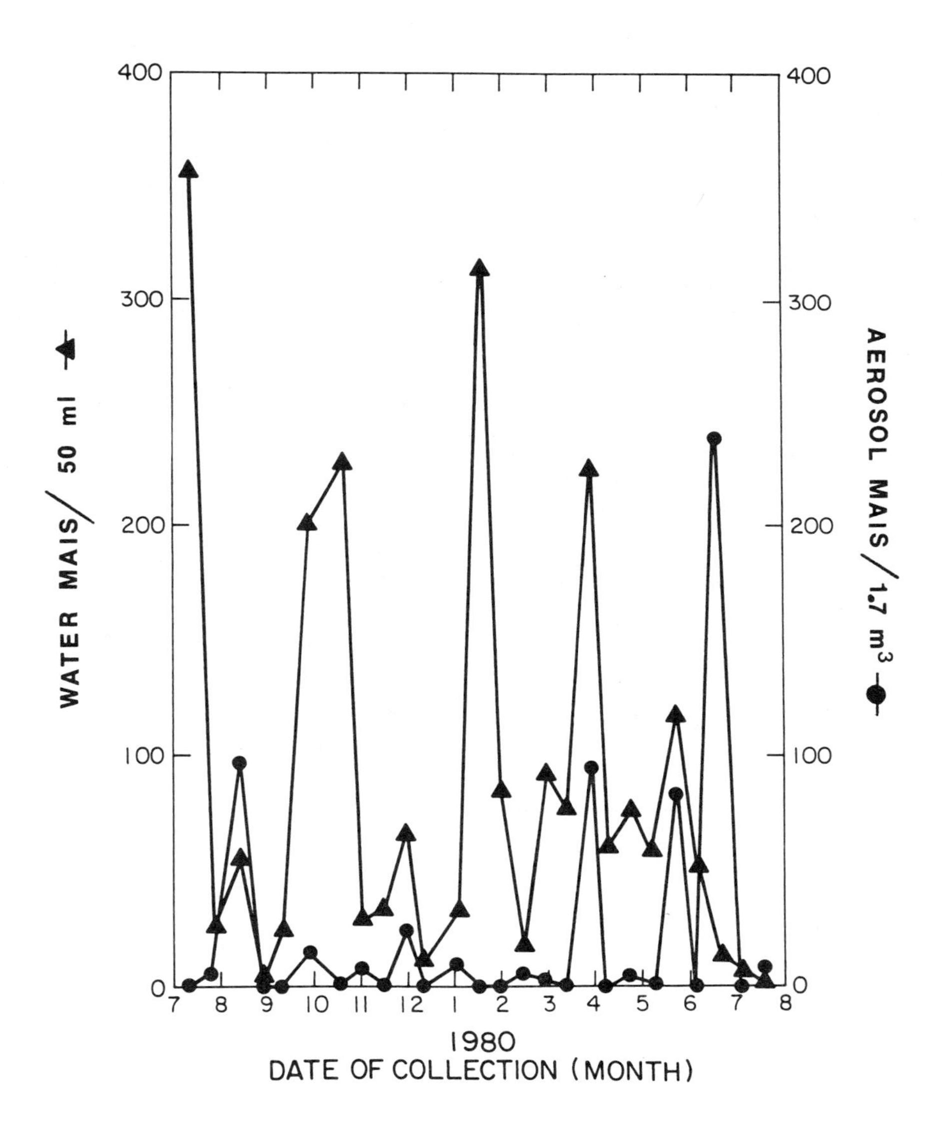

Though MAIS organisms were associated with particles of a size able to reach the human lung (≤ 5μ; stages 3, 4, 5 and 6), the majority were associated with larger particles (≥ 5μ; stages 1 and 2). On May 22, 1980, the size of MAIS-associated particles was smaller than those collected on the other days (Table 1). The size distributions of MAIS-associated particles collected at the James River (2 collections) and at Byrd Park were similar (Table 1).

M. terrae/*M. gordonae* and rapidly growing mycobacteria were associated with particles whose size range, with one exception, was similar to that of MAIS-associated particles. On August 16, 1979, the majority of rapidly growing mycobacteria were associated with particles of less than 1 μ (stage 6) collected at the south end of the Boulevard Bridge, while on June 19, 1980, 37 of 46 isolates of *M. terrae* and *M. gordonae* were associated with particles of greater than 8 μ (stage 1) collected at the north and south ends of the Boulevard Bridge.

Mycobacteria recovered from droplets ejected from the James River

Droplets ejected from natural waters by the bursting of air bubbles at the water surface result in the transfer of bacteria from water to air [3]. These droplets can be collected on the surface of bacteriologic medium and bacteria recovered as colony forming units. It is possible to determine the diameter and volume of such droplets using MgO-coated glass Petri dishes and thus calculate the concentration of bacteria in each droplet [6]. Table 2 shows the droplet diameter, the total volume of ejected droplets and the

TABLE 2 - Mycobacteria in ejected droplets collected from the James River

Date	Site[a]	Droplet Diameter[b]	Droplet Volume[c]	Numbers of Mycobacteria[d] MAIS	NON-MAIS	RAPID
11/6/79	H	220	190	0	0	1
1/17/80	H	170	30	1	3	0
3/13/80	H	220	230	0	0	3
4/24/80	H	240	340	0	0	0
5/11/80	H	210	150	0	0	0
6/5/80	H	160	20	1	0	0
7/31/80	H	180	10	0	0	0
8/14/80	S	140	5	0	7	4
10/15/80	S	175	130	1	20	37

[a] H = Hugenot Bridge; S = Sewage Treatment Plant.
[b] Droplet diameter expressed as microns (μ).
[c] Total droplet volume expressed as μ^3
[d] Number of mycobacteria recovered for the droplet volume indicated.

number and type of mycobacteria recovered (for that total volume) for collections made at the Hugenot Bridge and at the Sewage Treatment Plant. Few mycobacteria were recovered from ejected droplets collected at the Hugenot Bridge, while substantially more were isolated from droplets downstream at the Sewage Treatment Plant. Though the results were not statistically significant, ejected droplets collected at the Hugenot Bridge were bigger (200 ± 50 μ) than those collected at the Sewage Treatment Plant (150 ± 25 μ).

Mycobacteria in waters

Mycobacteria were recovered in high numbers from James River water samples (Figure 1 and Table 3). Though mycobacteria were recovered from water samples collected at all three sites, the numbers of each class were variable and the pattern of recovery differed (Table 3). MAIS organisms were recovered at approximately equal numbers from all three sites (Table 3). By contrast, highest numbers of *M. terrae*/*M. gordonae* (NON-MAIS) and rapidly growing mycobacteria (RAPID) were recovered from water samples collected downstream from the Richmond Municipal Sewage Treatment Facility (Table 3).

There was no correlation between aerosol and water numbers of MAIS organisms (Figure 1), *M. terrae*/*M. gordonae* and rapidly growing mycobacteria.

TABLE 3 -- Mycobacteria in water samples collected from the James River.

Collection Site	Numbers of Mycobacteria[a]		
	MAIS	NON-MAIS	RAPID
Sewage Treatment	28 ± 37	170 ± 306	89 ± 204
Boulevard Bridge	25 ± 37	29 ± 43	9 ± 12
Hugenot Bridge	34 ± 73	64 ± 237	9 ± 20

[a] Mean ± standard deviation of numbers of three groups per 150 ml water.

Comparison of mycobacteria isolated from water and aerosols

MAIS organisms can be separated into biovariants based upon the presence or absence of pigmentation (P) and urease (U) and catalase (C) activities [13,14]. *M. avium* and *M. intracellulare* lack pigment and urease and catalase activities (---). *M. scrofulaceum* is pigmented and has urease and catalase activities (+++). Significant numbers of MAIS intermediate types (e.g. +--, +-+) have been reported [13,14]. Table 4 displays percentages of the four major MAIS biovariants recovered from water and aerosol samples. Though one biovariant predominates (i.e. pigmented strains lacking urease and catalase activity), there is little difference between the distribution of these major biovariants in water and aerosol samples.

TABLE 4 - MAIS biovariants in water and aerosols.

Biovariant[a] P U C	Percent of MAIS organisms recovered Water	Aerosol
- - -	7	2
+ - -	60	80
+ + +	2	2
+ - +	14	13

[a] Presence (+) or absence (-) of pigmentation (P) and urease (U) and catalase (C) activity among isolates identified as MAIS [12]

DISCUSSION

These investigations establish that the slow growing and difficult-to-culture mycobacteria can be isolated from natural aerosols using the Andersen Cascade Sampler (Figure 1 and Table 1). Because of the instrument's ability to separate particles of different size [11] and the ability to incorporate an anti-fungal compound (i.e. malachite green) into the agar medium [4], few samples were lost due to fungal overgrowth. In addition, mycobacteria-containing droplets formed by bubbles bursting at the water surface were recovered on inverted bacteriological medium (Table 2) [4]. Further, inverted, MgO-coated glass Petri dishes were employed to determine the diameter of ejected droplets (here 100 - 200 μ). Consequently, the concentration of mycobacteria in ejected droplets can be calculated [6].

These investigations confirm our earlier report [4] of the presence of mycobacteria-containing aerosols along the James River in Richmond, Virginia. However, in contrast to that report we did not recover high numbers of MAIS organisms from aerosols collected at the Richmond Sewage Treatment Facility. Quite possibly, the wind direction was different for that collection than those reported for this study.

From the data in the Tables, the concentration of MAIS organisms was calculated. (1) Aerosol samples = 25.6 MAIS/m^3 on all stages and 1.4 MAIS/m^3 able to reach the alveloi. (2) Ejected droplets = 2,715 MAIS/ml. (3) James River Water Samples = 0.19 MAIS/ml. Based on an estimate that a human inhales an average of

6 m^3/day, 154 MAIS would be inhaled per day (8.3 reaching the alveoli; though not all all will implant). The low concentration in water samples and the fact that people rarely drink raw water, makes the James River an unlikely direct source of MAIS infection. An individual would have to drink 810 ml of water (almost 1 quart) in a day to take in as many organisms as inhaled in that same time period. The low numbers of MAIS organisms in James River waters contrasted with the high numbers in aerosols can be, in part, explained by the very high numbers in ejected droplets. We have earlier demonstrated such concentration of MAIS organisms in ejected droplets [6].

The large size of ejected droplets collected at both the Hugenot Bridge and Sewage Treatment Facility (Table 2) is consistent with the observation that mycobacteria collected from aerosols at those sites were associated with particles greater than 5 μ diameter (Table 1). However, it is also possible that there was another source for the mycobacterial-associated aerosols than the James River.

The observation that highest numbers of *M. terrae*/*M. gordonae* (NON-MAIS) and rapidly growing mycobacteria (RAPID) were found in water and ejected droplet samples collected from the Sewage Treatment Facility suggests that municipal sewage, rather than the James River itself, was the source of these two groups of mycobacteria. The low numbers of NON-MAIS and RAPID mycobacteria in aerosols collected at the Sewage Treatment Facility were unexpected, based on the high numbers in waters and ejected droplets. Quite possibly, these mycobacteria are poorly aerosolized or the droplets never dried to a size able to remain in air and precipitated in the humid Sewage Treatment Facility's aerosol. Alternatively, the organisms could not survive in aerosols.

Though MAIS organisms were recovered from aerosols, a number of yet unidentified factors must influence their numbers. The wide fluctuation in numbers in aerosols which did not correlate with numbers in James River waters (Figure 1), suggests that either: (1) there are other sources of MAIS in aerosols, (2) that water is the source and the aerosolization process selects a small subpopulation [6] or (3) both other sources and fractionation influence mycobacterial numbers in aerosols. Because highest MAIS numbers were recovered upstream of Richmond at the Hugenot Bridge, at least the River itself may be a source of these human pathogens in aerosols. That is consistent with the recovery of MAIS from other southeastern United States waters [3] and the fact that *MAIS* organisms are capable of growth in natural, unamended waters [15]. If there is another source of airborn MAIS, factors influencing aerosolization from that source affect all three mycobacterial groups, because high numbers of MAIS organisms in aerosols were correlated with high numbers of *M. terrae*/*M. gordonae* and rapidly growing mycobacteria in aerosols as well. Further, since the percent distribution of MAIS biovariants was the same for water and aerosol isolates (Table 3), the other source(s) of that group of pathogenic mycobacteria must have the same biovariants and in the same proportion.

An alternative explanation for the lack of correlation between numbers of mycobacteria in waters and aerosols (Figure 1) is that only a fraction of water-borne mycobacteria are aerosolized. MAIS strains differ widely in their ability to be aerosolized from water [6] and aerosols are only composed of those highly aerosolized strains. Quite possibly, there would exist a correlation between numbers of hydrophobic, water-borne mycobacteria and numbers in aerosols.

ACKNOWLEDGEMENTS

The research described was supported by awards from the National Institute of Allergy and Infectious Disease.

REFERENCES

[1] Riley, R.L., "Disease transmission and contagion control, "American Review of Respiratory Disease, Vol. 125, Koch Centennial Memorial, 1982, pp 16-19.

[2] Wolinsky, E., "Nontuberculous mycobacteria and associated diseases," American Review of Respiratory Disease, Vol. 119, 1979, pp 107-159.

[3] Falkinham, J.O., III, Parker, B.C. and Gruft, H., "Epidemiology of infection by nontuberculous mycobacteria. I. Geographic distribution in the eastern United States," American Review of Respiratory Disease, Vol. 121, 1980, pp 931-937.

[4] Wendt, S.L., George, K.L., Parker, B.C., Gruft, H. and Falkinham, J.O., III, "Epidemiology of infection by nontuberculous mycobacteria. III. Isolation of potentially pathogenic mycobacteria in aerosols," American Review of Respiratory Disease, Vol. 122, 1980, pp 259-263.

[5] Brooks, R.W., Parker, B.C., Gruft, H. and Falkinham, J.O., III, "Epidemiology of infection by nontuberculous mycobacteria. V. Numbers in eastern United States soils and correlation with soil characteristics," American Review of Respiratory Disease, Vol. 130, 1984, pp 630-633.

[6] Parker, B.C., Ford, M.A., Gruft, H. and Falkinham, J.O., III, "Epidemiology of infection by nontuberculous mycobacteria. IV. Preferential aerosolization of Mycobacterium intracellulare from natural waters," American Review of Respiratory Disease, Vol 128, 1983, pp 652-656.

[7] Meissner, P.S. and Falkinham, J.O., III, "Plasmid DNA profiles as epidemiological markers for clinical and environmental isolates of Mycobacterium avium, Mycobacterium intracellulare and Mycobacterium scrofulaceum," Journal of Infectious Diseases, Vol 153, 1986, pp 325-331.

[8] Fry, K.L., Meissner, P.S. and Falkinham, J.O., III, "Epidemiology of infection by nontuberculous mycobacteria. VI. Identification and use of epidemiologic markers for studies of Mycobacterium avium, M. intracellulare and M. scrofulaceum,"

American Review of Respiratory Disease, Vol. 134, 1986, pp 39-43.

[9] Blaser, M.J. and Cohn, D.L., "Opportunistic infections in patients with AIDS: clues to the epidemiology of AIDS and the relative virulence of pathogens," Reviews of Infectious Disease, Vol. 8, 1986, pp 21-30.

[10] duMoulin, G.C., Stottmeier, K.D., Pelletier, P.A., Tsang, A.Y. and Hedley-Whyte, J., "Concentration of Mycobacterium avium by hospital hot water systems," Journal of the American Medical Association, Vol. 260, 1988, pp 1599-1601.

[11] Andersen, A.A., New sampler for the collection, sizing, and enumeration of viable airborne particles," Journal of Bacteriology, Vol 76, 1958, pp 471-484.

[12] David, H.L., Bacteriology of the Mycobacterioses, Centers for Disease Control, Atlanta, 1976.

[13] Hawkins, J.E., "Scotochromogenic mycobacteria which appear intermediate between Mycobacterium avium-M. intracellulare and Mycobacterium scrofulaceum," American Review of Respiratory Disease, Vol 116, 1977, pp 963-964.

[14] Portaels, F., "Difficulties encountered in identification of M. avium-M. intracellulare, M. scrofulaceum and related strains,"American Review of Respiratory Disease, Vol 118, 1978, pp 969.

[15] George, K.L., Parker, B.C., Gruft, H. and Falkinham, J.O., III, "Epidemiology of infection by nontuberculous mycobacteria. II. Growth and survival in natural waters," American Review of Respiratory Disease, Vol. 122, 1980, pp 89-94.

QUESTIONS:

Question 1. Please describe the culture medium and incubation technique for Mycobacterium avium and M. tuberculosis. What is the best sampling technique for each species?

Answer. A standard mycobacterial culture medium, Middlebrook 7H10 containing 0.5% (vol/vol) glycerol and 0.0025% (wt/vol) malachite green has been employed for recovery of M. avium from aerosols. Following exposure of plates in the Andersen sampler, the medium is incubated at 37°C for 3-4 weeks at which point colonies can be tested to whether they are mycobacteria (acid-fast staining positive). If acid-fast, each is purified identified by biochemical tests.

We have had no experience in attempting to isolate M. tuberculosis from aerosols. Presumably, the same medium could be employed with incubation in 5% CO_2 for up to 8 weeks.

Question 2. Please comment on ingestion as a source of M. avium infection in AIDS.

Answer. Because M. avium infection is disseminated, not exclusively pulmonary, and M. avium can be recovered from feces in AIDS patients, it is thought that the major route of infection in AIDS patients is through gastro-intestinal tissue. Thus, ingestion is a possible route of M. avium infection in AIDS patients.

Question 3. What are the health affects (symptoms, course, etc.) of M. avium infection? Rate of infection? Are they a problem in office buildings, schools, homes? How to control them in those environments and sources?

Answer. In non-AIDS patients, the symptoms of M. avium infection are similar to tuberculosis, though it requires treatment with different and often multiple antibiotics. In AIDS patients it presents as a disseminated infection, frequently recovered from many different tissues and patient samples. Until the AIDS epidemic, it was estimated there were approximately 2,000 cases of M. avium infection annually in the United States. Estimates of the percentage of M. avium infections among AIDS patients range from 25% to 100%. To date, there has been no evidence of outbreaks of M. avium infection and pulmonary disease in office buildings, schools or public buildings (as has occurred with Legionella sp.). It has been suggested that because M. avium has been recovered from hospital hot water supply systems, they may serve as a source of infection (especially for AIDS patients). Obtaining retrospective evidence of M. avium infection will be difficult in the southeastern United States because over 50% of individuals demonstrate evidence of previous infection (i.e. skin test positive). Control of MAIS organisms at their sources (i.e. natural waters and soils) would be difficult, because of the nature of those sources and the relative resistance of mycobacteria to antibacterial agents. Disinfection of water supplies is possible, however mycobacteria are more resistant to chlorine than other bacteria.

Question 4. (1) How exactly is hydrophobicity measured? (2) Please provide citations for: jet drop bubble mechanism and the table of aerosolization of bacteria. (3) Please elucidate on problem of viable but non-culturable isolates. (4) How was surface charge measured? (5) Please list factors affecting aerosolization.

Answer. (1) Hydrophobicity can be measured by the adherence of bacteria to hexadecane (Rosenberg, FEMS Microbiol. Lttrs. 22:289, 1984). (2) The jet drop bubble mechanism is outlined in Weber, et al. (Limnol. Oceanogr. 28:101, 1983) and a table showing the aerosolization of bacteria given in Parker, et al. (Am. Rev. Respir. Dis. 128:652, 1983). (3) The number of mycobacterial cells which appear viable based upon substrate reduction or radioisotope uptake do not equal the number of colony forming units (CFUs) recovered on bacteriologic medium. CFUs can be increased by including pyruvate in medium to scavenge toxic oxygen radicals and incubating medium at low oxygen tension, suggesting such cells are inhibited from colony formation. (4) Surface charge was measured in a cytopherometer (George, et al. Microbios 45:199, 1986). (5) MAIS aerosolization is influenced by differences in (i) species, (ii) strain, (iii) membrane protein composition, (iv) growth medium, (v) growth stage, (vi) cell aggregation, (vii) water salinity and (viii) cell hydrophobicity.

Question 5. Since *Mycobacterium avium* aggregates, don't you think aerosol sampling with an impactor-type sampler drastically underestimates the number of airborne organisms in a volume of air?

Answer. Yes, microbial aggregation can lead to underestimates of numbers in aerosols. We have found that spreading the collected droplets using a sterile glass rod can result in significantly higher colony numbers, suggesting that the sheer forces generated by the glass rod can disperse the aggregates (Parker, et al. Am. Rev. Respir. Dis. 128:652, 1983).

Question 6. What was the ventilation strategy in the operating theater where the first case study (tube in throat) of *M. tuberculosis* outbreak?

Answer. Unfortunately, there was no air flow from the outer hall into the examining room and emergency room personnel were unaware of the patient's tuberculosis.

Question 7. How close is the commercial use of PCR (polymerase chain reaction) technology as is applies to mycobacteria? Are there any proposed ASTM methods directed at the use of PCR in analyzing any environmental agents and DNA probe development?

Answer. Development of DNA probes and PCR technology to determine whether mycobacteria are present in either an environmental or patient sample is well underway. This is especially critical for mycobacteria because their slow growth prevents rapid diagnosis of infection. I am unaware of any development of standards for the employment of such probes with or without PCR technology.

P. JULIAN L. DENNIS

AN UNNECESSARY RISK: LEGIONNAIRES' DISEASE

REFERENCE: Dennis, P.J.L., 'An Unnecssary Risk: Legionnaires' Disease', Biological Contaminants in Indoor Environments, ASTM STP 1071, Philip R Morey, James C Feeley, Sr., James A Otten, editors, American Society for Testing and Materials, Philadelphia, 1990.

ABSTRACT: Common aquatic bacteria like *Legionella pneumophila* are able to colonize man-made water systems and poorly maintained systems or those that are seldom used. These water systems often also provide the means for aerosolizing the organism. Construction materials and compounds leached from the by-products of other organisms can be utilized by aquatic bacteria and thus aid their growth. Keeping water systems clean and well serviced, keeping hot water at or above 50°C and cold water below 20°C and additionally in cooling towers maintaining the required levels of biocide, will reduce or prevent the growth of legionella. To be certain that the control measures are successful, microbiological and chemical monitoring should be done. The results of this monitoring, as well as the maintenance work undertaken, should be kept in a log so that failures in treatment can be quickly detected and remedial action taken before any risk of infection arises. Adhering to these simple guidelines will not only significantly reduce the risk of infection, but will also provide systems that are more efficient and cheaper to run.

KEYWORDS: Legionella, growth temperature, hot water, maintenance, aerozolisation, control.

Dr P. J. L. Dennis is Microbiology Manager, Thames Water Utilities, England.

INTRODUCTION

In 1976 a new bacterium *Legionella pneumophila* [1] was isolated from American Legion attendees who had developed pneumonia following attendance at a convention in Philadelphia [2]. The organism and the disease it produces (Legionnaires' disease) are, however, not new. The first recorded isolation of a bacterium like *L. pneumophila* was made in 1947 [3] and the first well-documented outbreak of disease occurred in 1957 [4].

Clinically, the illness is characterized by early symptoms of malaise, muscle aches and a slight headache. Shortly after there is a rapid rise in fever associated with shaking chills. Chest pains often without a productive cough are common. Dysponsea (difficulty in breathing), abdominal pain and gastrointestinal symptoms can also occur [3]. Less than 5% of those exposed appear to develop the illness, in 10-15% of these the illness is fatal.

Pontiac fever [5] is a self-limiting influenza like non-pneumonia form of the disease which has been related to the inhalation of legionellas. The incubation period is short (usually 36 - 48 hours) and of those exposed to the aerosol 95% will become ill. The illness resolves spontaneously in 2 - 5 days.

The epidemiological investigation of the outbreak of pneumonia in Philadelphia suggested that infection was due to inhalation of the organism[2] and later epidemiological findings implicated cooling towers [6] and the water distribution systems within large buildings [7] as sources and means of airborne spread.

Aerosols containing legionella can be produced by showers [8], humidifiers[9], and cooling towers. _L. pneumophila_ survives well in aerosols[10] , in comparison with other bacteria. Although the growth phase[10] and substances aerosolized with the organism [11] can affect its survival, many interrelated environmental factors may influence aerosol survival and hence infectivity. The critical factors associated with cooling towers probably differ from those in the more enclosed environment of a shower room.

Large volumes of bacterial sludge and slime containing high numbers of legionella have been responsible for at least one well documented outbreak of Legionnaires' disease [12] which happened after the sludge from a calorifier had been discharged through the plumbing system. It is, therefore, likely that the concentration of cells disseminated in aerosols, or present in the water in the system, is one of the most important factors in the sequence of events leading to infection.

Legionellas are aquatic organisms that have been isolated from a range of habitats, in particular from thermally polluted water and natural thermal ponds ranging in temperature between 20°C and 70°C. They have been isolated from streams, rivers and the shores of lakes [13] and ponds located in the blast zone of a volcano [14] . They are most commonly isolated from warmer waters (40-60°C)[13].

L. pneumophila has been found to grow with and derive essential nutrients from amoebae (growing intracellularly)[15], cyanobacteria [16] and flavobacteria [17]. They are gram-negative aerobic rods, 0.3-0.9 um in width and 2-20 um or more in length. They are not acid fast, and are motile by one, two or more polar flagella. L-cysteine -HCL and iron salts are required for growth.

It is inevitable that legionellas will enter and colonize man-made water systems as they form a part of the natural aquatic bacterial population of lakes and rivers from which we draw a lot of our water for domestic as well as industrial use. Aquatic bacteria may enter by surviving traditional water treatment or during repair and construction activities [18].

The degree of microbial growth within any water system depends on water temperature, redox potential, pH, the residence time of the water within the system, the concentration and persistence of any residual disinfectant, the type of water treatment process applied, the availability of organic or inorganic substrates and the presence of sediments or corrosion products [19].

It must be remembered that legionellas are only one member of a diverse population of microorganisms that find their way into and colonize water systems. The presence of aquatic bacteria in potable water is acceptable, there being no adverse health effects associated with their consumption. In the United Kingdom the Water Regulations [20] do not give permissible counts/ml at 22°C or 37°C but merely state that "There should be no significant increase in that normally observed". In cooling water systems total counts of bacteria of up to 10^{4}/ml at 22°C and 37°C are considered acceptable.

Most aquatic bacteria are harmless to humans, however some, like Pseudomonas aeruginosa are opportunistic pathogens that can give rise to serious illness, particularly in the elderly and those who are chronically ill or immuno-suppressed. These organisms colonize water systems and grow using the substrates (food) described above. This growth forms into biofilms or slime layers on the surfaces of pipes and tanks in contact with the water.

Slime layers can also promote corrosion of metal in systems that contain water of particular quality types, and therefore any control measures should be aimed not just at legionella but at the general microbial population. Controlling general microbial growth will thus not only reduce the risk of Legionnaires' disease but will also under some circumstances reduce or prevent corrosion as well.

HOT WATER PLUMBING SYSTEMS

Water for drinking and domestic purposes, in the UK, is derived in most cases from the public supply either by direct connection to the rising main or through properly constructed and protected storage cisterns. Water stored in cisterns other than those described above and which is not intended for consumption but is provided for domestic purpose (washing, heating etc.) or is conditioned (softened) before heating, that is derived from the public supply, naturally suffers some degree of deterioration in bacteriological and chemical quality and is in the UK considered to be no longer potable. Use of private supplies that are not adequately treated can have similar consequences. It is the colonization of these non-potable hot and cold water systems in large buildings like hospitals and hotels with legionella and its subsequent growth that is known to lead to outbreaks of Legionnaires' disease.

L. pneumophila has, however, also been isolated from the water system of buildings with no known associated legionellosis [21,22]. Temperatures between 30 and 54°C favour the growth of legionellas. In the USA and the UK the hot water system has been found to be the most common source[21,22]. Hot water tanks maintained at between 71 and 77°C were found not to contain the organism [22].

Legionellas gain sufficient nutrients to grow from water, from other organisms, and the materials used to construct water systems. It has been shown [23] that legionellas are able to grow in unsterile tap water and to continue to grow at 42°C when other organisms in the same water fail to do so. Natural rubber, sealing washers and gaskets have all been found to support the growth of legionellas [24,25]. In a hospital, only replacement of the rubber washers with washers made of suitable alternative materials finally eradicated the organism[24].

COOLING TOWERS

Problems with biofouling in cooling towers are well known and are caused mostly by the open nature of the systems which encourages addition of organic and inorganic contamination by entrainment in the cooling water, excessive aeration caused by the passage of water over the tower, and the constant addition of fresh water to make up for evaporative loss. This may be further compounded by the addition of inorganic nitrogen and phosphorous (corrosion and scale inhibitors added to the water) and process leaks as well as nutrients from construction materials dissolved in the water. Large amounts of slime impair efficiency and, as engineers well know, will incur high operating costs. To control this fouling, biocides and dosing regimes for their effective use have been developed.

Many small cooling towers are, however, used only intermittently, and maintenance may be inadequate, for example using auxillary towers [6] in which legionellas are able to survive and grow as part of the colonizing biofilm (slime).

In cooling systems that are not regularly cleaned or maintained, microbial sludge builds up in the reservoir of the tower and slime adheres to the wet surfaces so that large concentrations of bacteria, which may include legionellas, will accumulate.

MAINTENANCE AND MONITORING

All cooling systems require maintenance to prevent corrosion, scaling and biological fouling as described above. Biocides with or without biodispersants are used to control biological fouling. A number of biocides have been shown to kill legionella in laboratory experiments [26,27,28,29] but demonstration of the effect of the biocide on legionella in cooling systems or potable hot water systems remains to be done. If added on a regular basis, biocides will not only control general microbial growth, but also that of legionella, partly as a result of controlling other bacteria, algae, and protozoans, upon which the legionella may depend. Biocides can be added automatically to cooling systems with the aid of a pump and timer. The decreased handling by this method is preferable to the manual addition of the biocide with the aid of a bucket. Many factors can influence the choice of biocide and the concentration used. These determinations are most appropiately made by the water treatment company. The efficiency of any biocide is severely impaired by accumulation of debris and slime. Thus, to maximize biocide efficacy, cooling water systems must be kept as clean as possible.

Similarly, hot and cold water systems should be regularly cleaned and because it is inappropriate in most circumstances to add biocides or disinfectants to these systems, they should be operated in such a way that microbial growth is minimized. The presence of a heterotrophic population in potable water distribution systems has been documented[30]. Although, in this study [30] no attempt was made to culture legionella, the isolation of the same legionella serotype from a patient and the home water supply has been demonstrated[31].

Detailed procedures for operating and maintaining domestic and cooling water systems are outlined in a number of publications[32,33]. In general these documents emphasize that visual, chemical, and microbiological monitoring of cooling systems must be part of the routine operation of an efficient system. Monitoring will reveal whether or not the system is clean and if the biocide has been reaching the system at the appropriate concentrations.

SAMPLE COLLECTION

Water Services and Cooling Towers.

When investigating the water services within a building to determine whether they are colonized with legionellas, it is essential to prepare or obtain a simple schematic diagram of these services. The following features, if present, should be noted:

1. The location of the incoming supply and/or private source;
2. The location of cisterns, booster vessels and pumps;
3. The location of calorifiers (water heaters);
4. The type of fittings used in the system, e.g. taps, showers, valves, and the material from which the pipework is contructed;
5. Whether or not cooling towers or heating circuits are present;
6. Whether air conditioning systems and humidifiers are present within the building.

The route of the service should be traced from the point of entry of the water supply. The condition of pipes, the jointing methods used, the presence of lagging, sources of heat, and the standard of protection afforded cisterns should be noted on the diagram. A careful note should also be made of disconnected fittings, 'dead-ends', and cross-connections with other services. Having identified these sites, water samples should be taken from:

1. The incoming supply;
2. Cisterns and calorifiers;
3. An outlet close to, but downstream of, each cistern and calorifier;
4. The distant point of each service;
5. The water entering and leaving any fitting under particular suspicion.

Samples should be collected in polyolefin containers which if reused can be cleaned either by rinsing with distilled water or mains tap water followed by pasteurization with flowing hot (>70°C) water, steam for 15 minutes or autoclaving at 121°C for 15 minutes. Smaller wide necked polyolefin containers or glass can be used to collect sediments, deposits or slime.

It is important that the outlet is not flushed before samples are collected. The external surface and rim of the outlet being sampled should be clean and free from deposits.

1. A 200ml - 1L 'preflush' sample of water is collected;
2. The residual chlorine levels are measured (if required);
3. A thermometer is placed in the water flow and the outlet is

allowed to discharge until the temperature stabilizes. The initial and final temperatures, and the time taken to reach the latter, are recorded;

4. A 200ml - 1L 'after-flush' sample of water is collected.

The initial sample is intended to indicate the level of comtamination at the sample point/fitting and the final sample should reveal the quality of the water being supplied to the fitting.

Cooling towers should be sampled in the following manner. Water samples should be taken from the incoming supply to the tower, either from the header tank or the ball valve located in the tower pond. Samples should also be taken of the pond water furthest from the water make-up, and of the water returning from the circulation system at the point of entry to the tower.

For both cooling towers and building water services, sludge, slimes or sediments can be collected particularly where accumulations occur. Swabs of shower heads, pipes and faucets can also be taken and rehydrated in sterile distilled water or water taken from the sampling site.

Air Sampling

As a routine monitoring tool or risk assessement method, air sampling is of limited valve. It is likely that aerosols of legionella are not continuously generated but are more probably produced as a bolus when faucets or showers are turned on [24] or after the system has been disturbed [34]. Air sampling has however been of value in research where the production of aerosols and survival of legionella in them has been studied [35]. During these studies it was found that collection into liquid using impingers or cyclones was far superior to impaction methods (unpublished), with distilled water or a low ionic stength saline (Pages saline) being the collecting fluids of choice [10].

ANALYSIS

Samples should be allowed to attain ambient temperature during transportation. They should not be cooled but must be protected from sunlight or from other external heat sources. They should be delivered to the laboratory as soon as possible and preferably within two days. Ideally, the samples should be stored at room temperature (20 $\pm$ 5°C) and should be processed within 2 days. It has been found that samples can be stored for up to ten days without affecting the count of legionella. However, this can vary greatly according to the origin of the sample.

Methods for the isolation of legionellas are now well established [36]. However, quality control of culture media is of particular importance. Laboratory adapted strains, that is strains of legionella that have been grown on laboratory media for at least ten subcultures, should not be used when testing the ability of new batches of media to support the growth of legionellas. It is preferable to use freshly isolated strains or waters known to contain legionella. Using the latter, the number of legionella isolated (from a well mixed sample) and the colonial morphology of the isolates on the new medium can be compared with the previous batch. A simple statistical assessent can then be made and the quality of the medium determined. Screening samples using direct (DFA) or indirect (IFA) fluorescent antibody techniques have proved useful as a screening tool although these techniques will identify dead or moribund bacteria. Similarly the Legionella Rapid Assay (LRA, Boots Microcheck, The Boots Company, Nottingham) can be used as a screening method to supplement culture.

When investigating outbreaks of Legionaires' disease, it is essential that isolates are stored and compared with clinical isolates serologically using poly and monoclonal antibodies or Restriction Fragment Lenght Polymorphisms (RFLP's) of the microbial DNA, [37]. Using these techniques, the source of infection can be traced and remedial action taken.

Where there is no associated illness, assessment of risk cannot be made on single samples, although counts of greater than 1/ml in the U.K are generally thought unacceptable. Samples should be taken regularly and analytical results recorded so that results may be compared and sudden changes in count identified.

Monitoring will reveal whether cleaning has been successful and whether biocides (in cooling systems) are working properly, and provide confidence in the running and maintenance procedures. Even when an effective mode of operation and treatment package has been established, recirculating water systems can be subject to rapid alterations in water quality for reasons outside the direct control of the plant operators. To be effective, monitoring must be undertaken on a regular basis so that a history or standard can be established for every system.

Any changes from this standard (base line) can be immediately detected and remedial measures undertaken. The parameters measured and sampling frequency can vary and will depend on the type of system and the way it is used.

The Second Report of the Committee of Inquiry into the Outbreak of Legionnaire's disease in Stafford[32] in April 1985 recommended that routine monitoring for legionella should be undertaken only as part of a structured monitoring regime for the control of water quality. It is thus important that owners or engineers responsible for the safe running and maintenance of domestic and cooling water systems establish effective monitoring programmes, and when doing this seek expert assistance.

The views expressed in this paper are those of the author and not necessarily those of Thames Water Utilities Ltd.

REFERENCES

[1] McDade JE, Sheppard CC, Fraser DW, Tsai TR, Redus MA, Dowdle WR and the Laboratory Investigation Team. Legionnaires' disease: isolation of a bacterium and demonstration of its role in respiratory disease. New England Journal of Medicine;297, 1977, pp. 1192-203.

[2] Fraser DW, Tsai RT, Orenstein W, Parkin WE, Beecham PHJ, Sharrer RG, Harris J, Mallison GF, Martin SM, McDade JE, Sheppard CC, Brachman PS and the Field Investigation Team. Legionnaires' disease: description of an epidemic of pneumonia. New England Journal of Medicine;297, 1977, pp. 1189-97.

[3] McDade JE, Brenner Dj, Bozeman FM, In: Balows A, Fraser DW, eds. International Symposium on Legionnaires' Disease. Annals of Internal Medicine;90, pp. 659-61.

[4] Osterholm MT, Chin TY, Osbourne DO, Dull HB, Deon AG, Fraser DW. A 1957 outbreak of Legionnaires' disease associated with a meat packing plant. American Journal of Epidemiology;117, 1983, pp. 60-7.

[5] Glick T.H., Gregg, M.B., Berman, B., Mallison, G., Rhodes, W.W. and Kassanoff, I Pontiac Fever! An epidemic of unknown etiology in a health department:I. Clinical and epidemiological aspects. American Journal of Epidemiology;107, 1978, pp. 149-60.

[6] Dondero TJ, Rendtorff RC, Mallison GF, Weeks RM, Ley JS, Wong EM, Schaffner W. An outbreak of Legionnaires' Disease associated with a contaminated air-conditioning cooling tower. New England Journal of Medicine;302, 1980, pp. 365-70.

[7] Tobin J O'H, Beare J, Dunnill MS, Fisher-Hock SP, French M, Mitchell RG, Morris PJ, Muers MF. Legionnaires' disease in a transplant unit: isolation of the causative agent from shower baths. Lancet;ii;1980, pp. 118-21.

[8] Dennis PJ, Wright AE, Rutter DA, Death JE, Jones BPC. Legionella pneumophila in aerosols from shower baths. Journal of Hygiene;93, 1984 pp. 349-55.

[9] Zuraleff TJ, Yu VL, Shonnard JW, Rihs JD, Best M.

Legionella pneumophila contamination of a hospital humidifier. Demonstration of aerosol transmission and subsequent subclinical infections in guinea-pigs. Americal Review of Respiratory Diseases;128, 1983, pp. 657-61.

[10] Hambleton P, Broster MG, Dennis PJ, Henstridge R, FitzGeorge R, Conlan JW. Survival of Virulent Legionella pneumophila in aerosols. Journal of Hygiene;90, 1983, pp. 451-60.

[11] Berendt RF. Influence of blue-green algae (cyanobacteria) on survival of Legionella pneumophila in aerosols. Infection and Immunity;32, 1982, pp. 690-92.

[12] Fisher-Hock SP, Smith MG, Colbourne JS. Legionella pneumophila in a hospital hot water cylinder. Lancet;i, 1984, pp. 1073.

[13] Fliermans CB, Cherry WB, Orrison LH, Smith SJ, Tison DL, Pope DH. Ecological distribution of Legionella pneumophila. Applied and Environmental Microbiology;41, 1981, pp. 9-16.

[14] Tison DL, Baross JA and Seidler RJ. Legionella in aquatic habitats in the Mount Saint Helens blast zone. Current Microbiology;9, 1983, pp. 345-8.

[15] Rowbotham TJ. Preliminary report on the pathogenicity of Legionella pneumophila for fresh water and soil amobae. Journal of Clinical Pathology;33, 1980, pp. 1179-83.

[16] Tison DL, Pope DH and Fliermans CB. Utilisation of algal extracellular products by Legionella pneumophila Abstracts of the annual meting of the American Society for Microbiology, 1980, pp. 91.

[17] Wadowsky RM, Yee RB. Satellite growth of Legionella pneumophila with environmental isolates of Flavobacterium breve. Applied and Environmental Microbiology;46, 1983, pp. 1147-9.

[18] Hutchinson M and Ridgway JW. Microbiological aspects of drinking water supplies. In: Skinner FA, Shewan JM, eds. Aquatic Microbiology Symposium Series 6: Academic Press, 1977, pp. 179-218.

[19] Colbourne JS. Materials usage and their effects on the microbiological quality of water supplies. In: Microbial Aspects of Water Management. Society of Bacteriology Symposium Series, 1986, pp. 16.

[20] The Water Supply (Water Quality) Regulations 1989.

Statutory Instruments 1989 No. 1147: HMSO.

[21] Dennis PJ, Taylor JA, FitzGeorge RB, Bartlett CLR, Barrow GI. Legionella pneumophila in water plumbing systems. Lancet;i, 1982, pp. 949-51.

[22] Wadowsky RM, Yee RB, Mezmar L, Wing EJ, Dowling JN. Hot water systems as sources of Legionella pneumophila in hospital and non-hospital plumbing fixtures. Applied and Environmental Microbiology;43, 1982, pp. 1104-10.

[23] Yee RB, Wadowsky RM. Multiplication of Legionella pneumophila in unsterile tap water. Applied and Environmental Microbiology;43, 1982, pp.1330-4.

[24] Colbourne JS, Pratt DJ, Smith MG, Fisher-Hock SP, Harper D. Water fittings as sources of Legionella pneumophila in a hospital plumbing system. Lancet;i, 1984, pp. 210-13.

[25] Schofield GM and Wright AE. Survival of Legionella pneumophila in a model hot water system. Journal of General Microbiology;130, 1984, pp. 1751-6.

[26] England A.C., Fraser D.W., Mallison G.F., Mackel D.L., Skaliy P., and Gorman G.W. Failure of Legionella pneumophila sensitivities to predict culture results from disinfection treated air-conditioning cooling towers. Applied and Environmental Microbiology;43, 1982, pp. 240-44.

[27] Grace R.D and Dewar N.E. Susceptibility of Legionella pneumophila to three cooling tower microbiocides. Applied and Environmental Microbiology;41, 1981, pp. 223-36.

[28] Skaliy P, Thompson, T.A. Groman G.W., Morris G.K., McEachearn H.W. and Mackell E.L. Laborating Studies of disinfectants against Legionella pneumophila. Applied and Environmental Microbiology;40, 1980, pp. 697-700.

[29] Soracco, R.J. Gill H.K., Fliermanns, C.B and Pope D.H. Susceptibilities of algae and Legionella pneumophila to cooling tower biocides. Applied and Environmental Microbiology;45, 1983, pp. 1254-60.

[30] Tuovinen O.H. and Hsu J.C.. Aerobic and Anaerobic Microorganisms in inbercles of the Columbus, Ohio, Water Distribution System. Applied and Environmental Microbiology;45, 1983, pp. 1254-60.

[31] Stout J.E., Yu V.L and Murala, P. Legionnaire's disease acquired within the homes of two patients. Journal of the American Medical Association; 257, 1987, pp. 1215-1221.

[32] Anon. Second Report of the Committee of Inquiry into the Outbreak of Legionnaire's Disease in Stafford in April 1985.
Comnd 256, London : HMSO, 1987.

[33] The Control of Legionellae in Health Care Premises: A Code of Practice: HMSO. ISBN 0 11 3212089.

[34] Shands, K. N., Ho, J. L., Meyer, R. D., Gorman, G. W., Edelstein, P. H., Mallison, G. F., Finegold, S. M., and Fraser, D. W., Potable water as a source of Legionnaires disease. Journal of the American Medical Association;253, 1985, pp. 1412-1416.

[35] Dennis, P. J., and Lee J. V. Differences in aerosol surviral between Pathogenic and non-pathogenic strain of Legionella pneumophila serogroup 1. Journal Applied Bacteriology;65, pp. 135-141.

[36] Dennis, P. J. L. Isolation of Legionellae from Environmental Specimens in A Laboratory manual for Legionella. Harrison, T. G. and Taylor, A. G., Wiley, 1988, pp. 31-44.

[37] Saunders N. A. and Harrison, T. G. The Application of Nucleic Acid Probes and Monoclonal Antibodies to the investigation of Legionella Infections in A Laboratory manual for Legionella. Harrison, T. G. and Taylor, A. G., eds, Wiley, 1988, pp. 137-153.

DISCUSSION

Will guidlines be established for legionella control, especially in hospital and nursing home settings? UK? International?

CLOSURE

Guildlines or Codes of Practice have already been established in the UK. E.g. 'The Control of Legionellae in Health Care Premises.' A Code of Practice. These Codes of Practice are continually reviewed and up dated in the light of our expanding knowledge of legionella's natural habit and its means of control.

DISCUSSION

In your experience:
1) What is the best temperature to ship bulk H_2O Samples for legionella?
2) What would reproduce the best data (for field studies), impactor samples (SAS, Anderson) or impingers? If

impingers, would the sampling media have best results if the Media were a Buffered Charcoal Yeast Extract (BCYE) broth W/antibiotics (then shipped to Lab) or buffered saline solution or distilled H_2O? If impactors, would you use the SAS with BCYE Rodac plates?

3) How should you ship plates from field to laboratory after sampling for legionella?
   Any comments on other bacterial sampling plate transport? (e.g. TSA plates, heterotrophic plates, blood agar, etc)

CLOSURE

1) Ambient temperature, keeping samples chilled can cause a reduction in count.
2) Impinger samplers (AGI type). The collecting fluids that we have found to be most effective are distilled water, Pages Saline or 1/40 Ringers Saline.
3) If impinger samples are plated on site, BCYE plates should have been pre-dried, the sample allowed to absorb and the plates transported in sealed canisters directly to the laboratory at ambient temperature.

DISCUSSION

1) How hot must water be to avoid legionella growth in hot water systems?
2) Has legionella been isolated from stagnant water pooling on roofs?

CLOSURE

1) Currently it is recommended that water should leave calorifiers in a pumped system at $60^{o}C$ to attain a temperature of at least $50^{o}C$ at the point of use.
2) Not to my knowledge.

DISCUSSION

I have seen algae growing in sun heated, ephemeral ponds near air intakes on the flat roofs of buildings. A potential source of indoor legionella?

CLOSURE

Although legionella may theoretically grow in these areas, the organism has to be aerosolized to enter the indoor environment. I would think it extremely unlikely that sufficent organisms would be aerosolized from these places to represent any hazard.

DISCUSSION

In non outbreak situations, which buildings should have their water systems cultured?

CLOSURE

In a building that has not been associated with an outbreak of Legionnaires' disease, other simpler more frequently used tests may be used (e.g. total viable counts) to estimate water quality. Legionella tests can be done as well but not on their own.

DISCUSSION

As an investigator, you come in contact with a variety of biological contaminants. Do you use, or recommend the use of protective equipment (masks, special clothing)?

CLOSURE

The use of protective equipment depends very much on the tasks being undertaken. The investigator should consider each occasion carefully. E.g. If sampling from a cooling tower; if pumps are switched off there is no need to wear a respirator. If the tower is operating at the time of sampling, a respirator may be considered appropriate protection.

DISCUSSION

Was the concentration of legionella in the cooling tower of the BBC outbreak higher than the cooling towers of neighbouring buildings? In other words, could one have <u>predicted</u> a high risk of transmitting Legionnaires disease.

CLOSURE

The number issue is complicated by the fact that some strains of <u>L. pnenmophila</u> sg. 1 appear to be more virulent than others. Therefore, although the higher the number, the greater the risk of inhaling a viable cell: lower numbers of some strains may cause infection compared to other strains.

DISCUSSION

Could Dr Dennis please describe the 'Microcheck' Legionella rapid assay test (a slide of which he showed) and how it should be used?

CLOSURE

The 'Microcheck' Legionella test is based on monoclonal antibody technology. It can be useful for monitoring but the results must be considered carefully in that it will generally detect cells that are dead or non-culturable as well as culturable bacteria.

DISCUSSION

1) What tests are currently available for testing water? Please describe each briefly stating their advantages and disadvantages.
2) What tests do you forsee for the future that could be used easily in building investigations?

CLOSURE

1) Currently we have culture, IFA or DFA detection of cells using polyclonal or monoclonal antibodies, and some detection kits based on ELISA type assays.
   Culture is the most widely used technique although it is slow, taking at least 3 days. It may not detect cells that are non-culturable or sublethally damaged cells. The microscopic methods of detection are quicker but tend to be more demanding on analysts and will detect cells that may be dead.
   Cross reactions with bacteria of different genera can also be a problem giving false positive results. Detection kits show great promise but have not been properly evaluated. They may have their greatest use for making building investigations easier.

Eugene C. Cole

THE CHLAMYDIAE: INFECTIOUS AEROSOLS IN INDOOR ENVIRONMENTS

REFERENCE: Cole, E. C., "The CHLAMYDIAE: Infectious Aerosols in Indoor Environments," Biological Contaminants in Indoor Environments, ASTM STP 1071, P. R. Morey, J. C. Feeley, Sr. and J. A. Otten, Eds., American Society for Testing and Materials, Philadelphia, 1990.

ABSTRACT: The Chlamydia bacteria consist of three species, *C. trachomatis*, *C. psittaci*, and *C. pneumoniae*. Of greatest concern as a contaminant of indoor air is *C. psittaci*, the causative agent of psittacosis, a reportable disease. Infectious elementary bodies are spherical, 0.25-0.3 μm in diameter, and easily aerosolized by infected parrots, parakeets, pigeons, cockatiels, turkeys, and other bird reservoirs. Most cases occur in pet bird owners or turkey processors. Exposure occurs as dried droppings or eye and nostril secretions of infected birds are inhaled by a susceptible host in an enclosed space. Symptoms include fever, pneumonia, cough, headache, and weakness. The illness may be self-limiting or treated with antibiotics. Fatalities are usually restricted to the elderly. Laboratory infections may occur. Hospital cases should be isolated in rooms with controlled airflow. For bioaerosol monitoring, the all-glass impinger (AGI-30) and membrane filter are recommended. Two impingers with sucrose-phosphate-glutamate (SPG) medium are used at each site for 30 mins, after which samples are processed for direct fluorescent antibody (DFA) testing and inoculation of tissue cell lines. Air monitoring cassettes with detergent-free membrane filters may be used. Collected aerosols are eluted into SPG and concentrated for DFA or culture. Monitoring personnel should wear respirators with HEPA cartridges. Environmental surfaces may be decontaminated with phenolic disinfectants.

KEYWORDS: chlamydia, psittacosis, parrot fever, ornithosis

Dr. Cole is a senior research microbiologist at Research Triangle Institute, P.O. Box 12194, Research Triangle Park, NC 27709.

CHARACTERISTICS OF AGENT

Classification

Members of the genus Chlamydia consist of procaryotic organisms that parasitize eucaryotic cells. The order Chlamydiales contains one family, Chlamydiaceae, and one genus, Chlamydia, which contains three species, C. trachomatis, C. psittaci, and C. pneumoniae.

Although described some 80 years ago, it was not until 1930 that C. psittaci was successfully isolated, and not until 1957 that isolation of C. trachomatis was achieved. Since chlamydiae develop only in the cytoplasm of eucaryotic cells, a property resembling the obligate parasitism of animal viruses, they were originally regarded as viruses. It eventually became obvious however, that the organisms are bacteria with unique characteristics since they: 1) reproduce by a mechanism of binary fission, 2) contain both deoxyribonucleic acid (DNA) and ribonucleic acid (RNA), 3) possess procaryotic ribosomes and synthesize their own proteins and nucleic acids, 4) possess cell wall material similar to gram-negative bacteria, 5) possess enzymatic activities, and 6) are susceptible to a variety of antibiotics.

Morphology, Biology, and Strain Variation

Members of the Chlamydia undergo a developmental cycle in which two distinct forms occur. The forms are particularly adapted for extracellular survival and intracellular growth. The extracellular forms, or elementary bodies (EB) are spherical and 0.25-0.3 μm in diameter. They are surrounded by a rigid trilaminar envelope similar in composition to those of other gram-negative bacteria [1]. The intracellular forms of chlamydia are referred to as the reticulate bodies (RB). The RB are much larger than the EB and differ from them in many ways (RNA/DNA ratio, presence of hemagglutinin, susceptibility to sonication and trypsin, etc.) [1]. RB are adapted only for intracellular function, do not survive extracellularly, and cannot infect new cells.

On contact with susceptible cells the EB attach to the surface of host cells and are actively ingested by endocytosis. In 6 to 8 hours after entering a host cell, the EB of C. psittaci are reorganized into metabolically active RB that divide continuously by binary fission. By 18 to 24 hours after infection, RB begin to convert to EB. After some 48 hours host cells begin to die and chlamydia are released by cell lysis [1]. The cycle for C. trachomatis is similar but extends for 72 to 96 hours [1].

C. psittaci and C. trachomatis

These two species of Chlamydia can be distinctly differentiated. C. trachomatis is sensitive to sulfonamides and forms inclusions containing glycogen, which allows detection of the agent by iodine staining. C. psittaci is resistant to sulfonamides and does not produce glycogen. C. trachomatis usually produces a single inclusion that displaces the nucleus, whereas C. psittaci inclusions rupture early and are distributed around the nucleus. Additionally, less than 10% homology exists between *C. trachomatis* DNA and *C. psittaci* DNA [2].

Chlamydiae have a complex antigenic structure. There is a genus antigen common to all chlamydiae, and there are a variety of species-specific antigens. For *C. psittaci*, a large number of serologically different determinants have been demonstrated. For *C. trachomatis*, there are 3 serologically distinct types of lymphogranuloma venereum (LGV) agents, and 12 serotypes of the trachoma-inclusion conjunctivitis-urethritis group [1]. Strains of *C. trachomatis* include all organisms causing trachoma in humans, inclusion conjunctivitis, LGV, and mouse pneumonitis. *C. psittaci* strains include human pneumonitis (Borg) or psittacosis, meningopneumonitis (Cal 10), feline pneumonitis, ovine pneumonitis, 6BC (parakeet), bovine chlamydial abortion (EBA-59-795), ovine chlamydial abortion, bovine encephalomyelitis, pigeon ornithosis, turkey ornithosis, sheep polyarthritis, and epizootic chlamydiosis [2].

TWAR Agent (C. pneumoniae)

A chlamydial agent distinctly different from *C. psittaci* and *C. trachomatis* may cause more human infection than those two species. The organism was isolated from the eye of a child in Taiwan (TW) in 1965, and from an acute respiratory (AR) infection in a university student in 1983 [3]. The TWAR organism differs from *C. trachomatis* in inclusion morphology, the lack of glycogen in inclusions, and failure to react with species-specific antibody for *C. trachomatis* [4]. TWAR inclusions fail to show the variable shape and growth around the cell nucleus that many *C. psittaci* strains do. Electron microscopy has shown that TWAR elementary bodies are unique from those of all other chlamydiae; and TWAR strains are also different from a number of *C. trachomatis* and *C. psittaci* strains in DNA homology, restriction endonuclease patterns, and the absence of plasmid DNA [4]. Because of these unique differences, the TWAR agent has been designated as a third species of *Chlamydia*, *C. pneumoniae* [5].

HUMAN HEALTH EFFECTS

C. trachomatis

Playing a major role in sexually transmitted disease, *C. trachomatis* is recognized as a major cause of nongonococcal urethritis and epididymitis in men, cervicitis and acute pelvic inflammatory disease in women, and inclusion conjunctivitis and a distinct form of interstitial pneumonia in infants. Except for the mouse pneumonitis agent, humans are the only known natural host for *C. trachomatis*. Infection with the organism occurs by contact with an infected individual in 3 ways: 1) direct contact with the eye secretions, 2) sexual intercourse, and 3) perinatal infection during passage through the birth canal, resulting in conjunctivitis and/or pneumonitis. Transmission of *C. trachomatis* by the aerosol route has been reported in laboratory workers following the sonication of infected cells in the open environment [6].

C. psittaci

Psittacosis is an infectious disease of birds and mammals and is transmissible to man. Human disease, which follows 4 to 15 days after exposure, may be mild, moderate or severe. Exposure usually occurs when the dried fecal droppings or eye and nostril secretions of birds containing the organism are aerosolized and inhaled by a susceptible host in an enclosed space. An attempt at defining an infectious dose was made by McGavran et al. who exposed 24 Rhesus monkeys to aerosols of a Borg strain of psittacosis [7]. The aerosol was disseminated into a chamber at 21-22°C with relative humidity ranging from 26-30%. The median diameter of the airborne particles was 1.0 to 1.5 μm. Anesthetized monkeys were exposed for 5 minutes. The calculated average inhaled dose of the agent was 4000-5000 $MICLD_{50}$ (the dose resulting in the death of 50% of a population of mice following intracerebral inoculation). The initial site of infection was in the respiratory bronchioles, spreading so as to result in lobular pneumonia. Maximum inflammatory reaction was reached by the 14th-16th days and was coincident with the onset of resolution. None of the infections resulted in death.

The most common symptoms of psittacosis in infected humans, as identified by the Centers for Disease Control (CDC) include fever, pneumonia, cough, headache, weakness or fatigue, chills, myalgia, chest pain, anorexia, nausea or vomiting, dyspnea, and diaphoresis [8]. Prodromal signs and symptoms may extend over 3-4 days, with anorexia and a slowly rising temperature. At the height of infection generalized severe headache and myalgia are dominant symptoms accompanied by a non-productive cough, abdominal distention and tenderness, and restlessness and insomnia [9]. Other specified clinical manifestations include sore throat, diarrhea, arthralgia, meningismus, rhinorrhea, photophobia, weight loss, dizziness, constipation, pleuritis, hepatitis, stiff neck, rash, hemoptysis, myopericarditis, splenomegaly, lymphadenopathy, polydipsia/polyuria, blurred vision, and hematuria. Fatal outcomes are usually restricted to the elderly and result from pulmonary insufficiency and overwhelming toxemia. Specific treatment is usually 2 g/day of tetracycline or erythromycin continued for 10-14 days after temperature returns to normal (1-2 weeks). An alternate therapy is 1 g/day for 21 days [5]. In some cases, prolonged therapy may be necessary. Diagnosis is confirmed serologically by the complement fixation (CF) test using acute and convalescent sera to observe a fourfold rise in antibody titer. A single sample titer of 1:32 or greater is also diagnostic. The microimmunofluorescence (MIF) test using chlamydial organisms grown in yolk sacs may also be used. The presence of circulating antibodies is not protective and will not prevent reinfection. The organism may also be isolated by culture in tissue cell lines and in mice.

Man to man transmission, presumably by the airborne route, has been documented and has resulted in disease clinically more severe than psittacosis acquired from birds [9]. While on average some 50-100 cases are reported each year nationwide, it is presumed that the number of actual cases is much higher, as those with mild cases may not seek medical attention or through misdiagnoses are subsequently treated with antibiotics.

C. pneumoniae

As with C. psittaci, the TWAR agent produces illness that may be mild, moderate or severe. Infection in adults includes mild to severe pneumonia, chronic bronchitis, pharyngitis, and sinusitis [5]. Clinical disease in teenagers and young adults may manifest as mild yet prolonged pneumonia, chronic bronchitis, and primary sinusitis or pharyngitis; while in children 6-12 years old infections can be milder with more asymptomatic cases; and young children (1-5 years of age) in tropical countries, may show serious lower respiratory tract infection [5]. Antibiotic therapy with tetracycline or erythromycin is the same as that for C. psittaci infections. Like C. psittaci, C. pneumoniae is not susceptible to sulfa drugs.

RESERVOIRS OF PSITTACOSIS

Birds

More than 130 species of birds, both domestic and wild, have been found to be infected with C. psittaci, and it is probable that all bird species can harbor and hence transmit the organism [10]. Psittacosis (from the Greek word for parrot, Psittakos) occurred rarely in the U.S. and Europe until the ownership of pet tropical birds became fashionable. A pandemic occurred in 1929-1930, when approximately 1,000 cases, including 200-300 deaths were attributed to psittacosis in various countries importing exotic birds from South America [11]. While psittacine birds (parrots, parakeets) are considered the major reservoir for C. psittaci, human cases have been associated with pigeons, sparrows, canaries, cockatiels, ducks, and turkeys. Conditions of overcrowding and dietary deficiencies may result in activation of inapparent infection, with transfer of the agent to previously unexposed and susceptible birds [11]. Fatal disease may occur in all bird populations. Owners of pet birds comprise approximately 50% of all cases [12]. Psittacosis can be considered an occupational hazard of pet shop employees, pigeon fanciers, zoo workers, veterinarians, and those involved in the handling of birds, particularly those processing domestic turkeys. Also, sporadic human cases from exposure to wild birds have been seen in hunters, people studying birds, or people trying to help sick birds. Psittacosis has been diagnosed in an arbovirus researcher who was bleeding cormorants in the field, and in a researcher studying vultures [13].

Mammals

Human infections with chlamydiae transmitted from nonhuman mammalian sources are rare yet have been documented: two laboratory infections (with a bovine and an ovine isolate); one naturally-occurring infection with the agent of feline pneumonitis; and a fatal case in which epidemiologic observations implicated cattle [11].

Man

Man to man transmission of psittacosis is infrequent, although a few instances of high transmission rates have been reported [11]. In such outbreaks, the secondary cases developed in persons who had attended untreated and severely ill index cases. Such risk appeared greatest for individuals present in the 48 hours before the death of the initial patients. *C. pneumoniae* is apparently a primary human pathogen transmitted from person to person without involvement of avian species or lower animals [5].

Ectoparasites

Arthropods, such as fleas or ticks, have not yet been shown to be true chlamydial vectors, although experiments indicate they may be mechanical vectors [11]. There is no evidence of multiplication in or transovarian transmission by ectoparasites.

Hospitals/Laboratories

Personnel involved in the direct care of psittacosis- or *C. pneumoniae*-infected patients with severe pneumonia are at risk of infection by the aerosol route. Outbreaks involving transmission to nurses are known. During a psittacosis epidemic in Louisiana in 1943, there were 8 deaths among 19 diagnosed infections in nursing attendants [14]. Laboratory personnel are at high risk of acquiring psittacosis when processing clinical specimens, if proper precautions and containment procedures are not followed. Such activities as mixing, shaking, blending, pipetting and centrifuging chlamydial specimens, can generate infectious aerosols. *C. psittaci* requires Biosafety Level 2 practices as recommended by the Centers for Disease Control and the National Institutes of Health, to include both animal and non-animal work [15]. Infections with chlamydiae are the fifth most commonly reported laboratory-associated bacterial infection [15]. An analysis of 3,921 lab-acquired infections for the period 1924-1974 was published by Pike [16]. Results showed 128 confirmed cases of chlamydial infection, 116 being psittacosis, with 10 deaths. This was the highest case-fatality rate of all groups of infectious agents. Twenty two (19%) of the psittacosis cases were identified as being contracted from known aerosol exposure. In 1987, Miller, Songer, and Sullivan published a 25-year review of lab-acquired infections at the National Animal Disease Center [17]. They reported that *Chlamydia* species were the 4th most common organism to which worker exposures occurred. Out of 12 chlamydial exposures, 4 resulted in infection.

SAMPLING METHODS

Rationale for Air Sampler Choice

Bioaerosol monitoring for *C. psittaci* has not been routinely employed in suspected outbreaks in enclosed facilities. Previously conducted laboratory aerobiological research, however, does provide guidance in selecting a suitable sampling method. Rosebury has reported the results of his aerosol chamber studies with three strains of C. psittaci: 6BC, Gleason, and Borg [18]. Using a glass impinger with 25 ml of collecting fluid (to which he added 4 drops of olive oil), he attempted recovery of C. psittaci aerosolized in a cloud chamber. He determined that extract broth (used for collecting viruses) and 10% sheep serum in distilled water were comparable in their collection of dry clouds of C. psittaci 6BC. Using the extract broth in expanded studies he realized recoveries of 2.8-38.4% when the 6BC strain was sprayed at relative humidities ranging from 35-70%.

One all-glass impinger, the AGI-30, is a popular bioaerosol sampler. It is one of only two samplers ever proposed as standards [19], is inexpensive, readily available, approximates the breathing rate of man and is recognized as having collection efficiency approaching 99%. The AGI-30 draws air at 12.5 L/min. The curvature of the inlet tube allows particles above respirable size to be impacted, while smaller particles are entrained in the airstream and move through a critical orifice where they approach sonic velocity and are impinged in the collection fluid. The turbulence of the fluid causes disruption of organism aggregates and allows for quantitation of discrete viable particles following dilution and plating. Used for monitoring for airborne particles containing C. psittaci EB, the AGI-30 lends itself to relatively easy processing in the laboratory as the collection fluid is centrifuged to concentrate the sample which can then be used to inoculate cell lines and/or perform diagnostic serology.

Aerosol impactors, such as the cascade sieve sampler or the slit sampler, have been used to collect viruses on solid media [20]. The new Andersen S-6 impactor uses only the 6th stage of the traditional 6-stage model as it collects airborne particles in the 0.65-1.1 μm range. Theoretically, it could be used to collect aerosolized C. psittaci although additional processing would be necessary in the laboratory to separate the organisms from the agar prior to concentration.

Membrane Filtration is a viable bioaerosol collection method that, like the AGI-30, can be used with relative ease for indoor monitoring, and is conveniently processed in the laboratory. Membrane filters and impingers have been found to be of equivalent efficiency in the collection of viruses [20]. Two and 3-stage 37 mm air monitoring cassettes fitted with sterile 0.45 μm membranes are commercially available and can be used at varying flow rates. Particles collected can be eluted from the membrane during laboratory processing. It is assumed that aerosolized C. psittaci EB will occur as associated particles larger than their individual 0.25-0.3 μm diameter.

Recommended Sampling Methods

Liquid impinger: The AGI-30 is available from Ace Glass Inc., Vineland, NJ. Two samplers should be used at each sampling site and operated for at least 30 minutes. Twenty ml of sucrose-phosphate-glutamate medium (SPG) with 0.02% streptomycin and 0.01% kanamycin is aseptically added to each sampler to serve as the collection fluid. To make SPG, combine 75 g sucrose, 0.52 g potassium dihydrogen phosphate (KH_2PO_4), 1.22 g sodium hydrogen phosphate (Na_2HPO_4), and 0.72 g glutamic acid. Add water to 1 liter. After sampling, recap intake and exhaust ports, place in crushed ice and transport to the laboratory. (Note: Normal saline will destroy C. psittaci EB and should never be used for collection).

Membrane filtration: Air monitoring cassettes, 37 mm, with 0.45 μm membrane filters are commercially available (e.g. Gelman Sciences, Ann Arbor, MI, Millipore Corporation, Bedford, MA). It is preferable that filters be detergent free if culture for chlamydiae is desired. It has been reported that fluids filtered through membranes containing a wetting agent (detergent) may be inhibitory to cell cultures [21]. As with the impingers, two cassettes are utilized at each sampling site. Airflow is adjusted to pull 2.0 L/min for 30 min. Following monitoring, inlets and outlets are recapped and the units transported to the laboratory.

Transport to the Laboratory

Liquid samples should be refrigerated if they are to be processed within 24 hours or frozen at -60 C in polypropylene freezer containers for longer storage. Membrane filter cassettes should be protected from temperature extremes and transported to the laboratory as soon as possible.

LABORATORY ANALYSIS

Direct Fluorescent Antibody Test

The most rapid assessment for the presence of airborne Chlamydia EB collected by liquid impinger or filter membrane involves the application of chlamydial antibody to fixed, concentrated samples. Impinger collection fluid is placed into 50 ml tubes and spun for 1 hour at 10-12,000 RPM in a Beckman, Sorvall or equivalent centrifuge. The supernatant is discarded. The pellet of debris is fixed with cold absolute methanol and applied to a slide or coverslip and allowed to air dry. The specimen is overlaid with fluorescein-conjugated antibody. The conjugate binds specifically with the antigen, if present, during incubation, with a wash step removing unbound antibody. Slides are then viewed under an epi-fluorescence microscope for typical apple-green fluorescent staining elementary bodies.

Filter membranes are removed from cassettes and placed into 30 ml of sterile SPG in 50 ml conical screw cap centrifuge tubes. Tubes are shaken by hand for 1 minute, the membranes removed and the samples processed as described.

Chlamydial Culture

Samples concentrated as described may also be inoculated into tissue culture. Cell lines that may be used include L-929 mouse fibroblast cells, HeLa 229 cells, and McCoy cells. Standard culture and isolation procedures must be followed closely. Briefly, the principle involves centrifugation of the inoculum onto the cell monolayer at approximately 2,800 X g for 1 hr, incubation of monolayers for 48-72 hr, and then staining with Giemsa stain to detect *C. psittaci* inclusions in cells [22].

Laboratory Safety

As *C. psittaci* is a highly infectious respiratory pathogen it is recognized as a significant hazard to laboratory personnel. Biosafety Level 2 recommendations must be followed [15]. All sample processing should be conducted in a Class I or II biological safety cabinet to include centrifugation, unless a specific aerosol-free centrifuge is used.

CASE STUDIES

Psittacosis Associated with Turkey Processing

As reported by the CDC [23], an outbreak of psittacosis occurred among employees of an Ohio turkey processing plant in July 1981. About 27 of the plant's approximately 80 employees were ill, 3 of them hospitalized. Turkeys slaughtered at the plant were the probable source of infection. The plant operates 40 hours/week, 10 months/year, processing only turkeys which are delivered by truck from various locations, and slaughtered and defeathered on the day of arrival. They are conveyed on a continuously moving line into the evisceration area, where deep tissues are exposed, the birds are inspected and trimmed, and edible organs are removed.

Most workers had an illness characterized by weakness, headache, fever, chills, and cough. To a lesser extent, they had photophobia, conjunctival suffusion, generalized joint pain, stomach cramps, and diarrhea. Eight workers showed evidence of pneumonia consistent with psittacosis upon x-ray. Paired serum samples from 27 workers were tested for complement-fixing antibodies. Of 15 workers who had illness compatible with psittacosis, 7 had a fourfold or greater rise in titer, and 5 had a titer of 1:16 or greater in at least one specimen. Of 12 workers who had not had a compatible illness, none had a significant titer change. Because most employees worked in various job stations in several departments on a given day, it was difficult to assess the relative importance of respiratory, skin, and conjunctival exposure. The attack rate by work department, however, was significantly higher for workers in the kill and evisceration areas. Additionally, there was no apparent correlation between degree of skin exposure and clinical psittacosis, suggesting that infections were the result of aerosol transmission or that multiple routes of exposure may have been involved.

In this outbreak, no environmental monitoring was done. Psittacosis was tentatively diagnosed based on the nature of the operation (turkey processing), the high attack rate, and the clinical disease. Confirmation of the agent's involvement was obtained serologically some weeks after the outbreak.

Psittacosis Outbreak Scenario

In a veterinary school building, many individuals have become ill, several with respiratory symptoms. The attack rate is about 20% and appears to vary according to HVAC zone. The building houses separated populations of birds and mammals and it is realized that the respiratory illness observed in the workers may be caused by any of a number of zoonotic disease agents transmissible to man.

Environmental monitoring personnel wearing respirators equipped with high efficiency filters begin collecting 30 min aerosol samples with AGI-30s containing SPG, in animal areas and offices where high attack rates were noted. Samplers are placed within the streams of exhaust air in the animal areas, and in the incoming flow of air in the offices. While conducting AGI monitoring each person wears a portable cassette sampler fitted with a 0.45 μm membrane filter operating at a flow rate of 2.0 L/min. While all animal populations are examined by veterinarians for diseased suspects, AGI and filter samples are processed by a diagnostic laboratory with chlamydial expertise, for both rapid direct fluorescent antibody in less than 24 hrs, and culture confirmation within 72 hrs.

ENVIRONMENTAL CONTROL

Quarantine

Because of latency of infection in many bird populations, quarantine of imported birds has historically not proven successful [11]. Present U.S. Department of Agriculture importation regulations require oral tetracycline treatment for 30 days in the United States for birds imported from all countries except Canada [8]. A majority of annual cases, however, are associated with domestically-raised parakeets and nonpsittacine birds such as pigeons and turkeys. Infected flocks of commercial turkeys are to be quarantined until diseased birds have been destroyed or adequately treated with tetracycline [24].

Sanitation/Chemical Germicides

After infected birds have been treated or destroyed, the rooms or areas where they had been housed must be thoroughly cleaned and disinfected with a phenolic compound [24]. Killed suspect birds should be immersed in 2% phenolic disinfectant, secured in a plastic bag, and shipped frozen to a competent laboratory for examination.

Hospital Infection Control

A psittacosis patient should be cared for in a single-occupancy room where masks are worn, hands are washed on entering and leaving, and articles contaminated with secretions are disinfected [8]. It is preferable that the room air pressure be negative to the public corridor and exhausted in such a manner so as to prevent dissemination of the airborne agent into other occupied areas.

Vaccines

No universally accepted efficacious vaccines for either animals or man are available.

Conclusion

Species of *Chlamydia* infectious for man by the aerosol route include *C. psittaci* and *C. pneumoniae*. Psittacosis is a reportable disease that occurs sporadically on an annual basis, usually as a result of human exposure to populations of infected pet birds or commercial flocks of domestic turkeys. *C. psittaci*, while species-specific for a variety of birds and mammals, is a significant human respiratory pathogen, and although treatable with antibiotics, it carries a risk of fatality to the elderly. A demonstrated antibody titer is not protective for subsequent reexposure, and endogenous reinfection is known to occur. *C. pneumoniae* (TWAR agent) is regarded as a species-specific respiratory pathogen of man. Infection may present as a mild to severe disease.

While chlamydial respiratory disease in man may be initially inferred from symptoms, attack rate, and recognized contact with host reservoirs, it is normally not confirmed for several days to a number of weeks. Aerosol monitoring techniques using the All-glass impinger (AGI-30) and membrane filter cassette have been described that may provide rapid evidence for the aerogenic spread of *Chlamydia psittaci*. Such information provides for the early administration of antimicrobial therapy in those infected and the immediate institution of appropriate environmental controls, both necessary for the quick termination of a recognized outbreak.

ACKNOWLEDGEMENT

The author acknowledges Dr. P. B. Wyrick of the University of North Carolina at Chapel Hill for providing valuable information used in the preparation of this paper. Partial support was provided by the Environmental Criteria and Assessment Office of the U.S. Environmental Protection Agency.

REFERENCES

[1] Manire, G. P. and Wyrick, P. B., "The Chlamydiae," in Infectious Diseases and Medical Microbiology. Braude, A. I., Davis, C. E., and Fierer, J., Eds., W. B. Saunders Company, Philadelphia, 1986, pp. 449-454.

[2] Becker, Y., "The Chlamydia: Molecular Biology of Procaryotic Obligate Parasites of Eucaryocytes," Microbiological Reviews, Vol. 42, No. 2, 1978, pp. 274-306.

[3] Grayston, J. T., Wang, S. P., Kuo, C. C., Mordhorst, C. H., Saikku, P., and Marrie, T. J., "Seroepidemiology with TWAR a New Group of Chlamydia psittaci," Abstracts of the ICAAC, 1984, p. 290.

[4] Marrie, T. J., Grayston, J. T., Wang, S. P., and Kuo, C. C., "Pneumonia associated with the TWAR strain of Chlamydia," Annals of Internal Medicine, Vol. 106, No. 4, April 1987, pp. 507-511.

[5] Grayston, J. T., "Chlamydia pneumoniae, Strain TWAR," Chest, Vol. 95, No. 3, March 1989, pp. 664-669.

[6] Peterson, K., "Chlamydia Infections Following Sonication," presented at XXV Biological Safety Conference, Boston, 1982.

[7] McGavran, M. H., Beard, C. W., Berendt, R. F., and Nakamura, R. M., "The Pathogenesis of Psittacosis: Serial Study of Rhesus Monkeys Exposed to a Small Particle Aerosol of the Borg Strain," American Journal of Pathology, Vol. 40, No. 6, June 1962, pp. 653-670.

[8] Center for Disease Control, "Psittacosis Surveillance, Annual Summary," Department of Health, Education, and Welfare, Atlanta, GA, 1978.

[9] Gregg, M. B., and Wehrle, P. F., "Psittacosis and Chlamydial Pneumonia of Infancy," in Communicable and Infectious Diseases, Wehrle, P. F., and Top, Sr., F. H., Eds., The C. V. Mosby Company, St. Louis, 1981, pp. 516-520.

[10] Schachter, J., "Chlamydia psittaci - Reemergence of a Forgotten Pathogen," The New England Journal of Medicine, Vol. 315, No. 3, 17 July 1986, pp. 189-191.

[11] Schachter, J., "Psittacosis," in Diseases Transmitted from Animals to Man, Hubbert, W. T., McCulloch, W. F., and Schnurrenberger, P. R., Charles C. Thomas, Springfield, IL, 1975, pp. 369-387.

[12] Schaffner, W., "Chlamydia psittaci (Psittacosis)," in Principles and Practices of Infectious Diseases, 2nd ed., Mandell, G. L., Douglas, R. G. Jr., and Bennett, J. E., Eds., John Wiley & Sons, New York, 1985, pp. 1061-1063.

[13] Schachter, J., and Dawson, C. R., Human Chlamydial Infections, PSG Publishing Company, Inc., Littleton, MA, 1978.

[14] Bennett, J. V., and Brachman, R. S., Eds., Hospital Infections, 2nd ed., Little, Brown and Company, Boston, 1986.

[15] Centers for Disease Control and National Institutes of Health, Biosafety in Microbiological and Biomedical laboratories, 2nd ed., U.S. Dept. of Health and Human Services, Washington, DC, 1988.

[16] Pike, R. M., "Laboratory-Associated Infections: Summary and Analysis of 3,921 Cases," Health Laboratory Science, Vol. 13, No. 2, April 1976, pp. 105-114.

[17] Miller, C. D., Songer, J. R., and Sullivan, J. F., "A Twenty-Five Year Review of Laboratory-Acquired Human Infections at the National Animal Disease Center," American Industrial Hygiene Association Journal, Vol. 48, No. 3, March 1987, pp. 271-275.

[18] Rosebury, T., Experimental Air-Borne Infection, The Williams and Wilkins Company, Baltimore, 1947.

[19] Fredkin, A., "Sampling of Microbiological Contaminants in Indoor Air," in Sampling and Calibration for Atmospheric Measurements, American Society for Testing and Materials, Philadelphia, 1987, pp. 66-77.

[20] Dimmick, R. L., Akers, A. B., Heckly, R. J., and Wolochow, H., Eds., An Introduction to Experimental Aerobiology, John Wiley & Sons, Inc., New York, 1969.

[21] Cahn, R. D., "Detergents in Membrane Filters," Science, Vol. 155, 13 January 1967, pp. 195-196.

[22] Schachter, J., "Chlamydiae (Psittacosis-Lymphogranuloma Venereum-Trachoma Group)," in Manual of Clinical Microbiology, 4th ed. Lennette, E. H., Balows, A., Hausler, Jr., W. J., and Shadomy, H. J., Eds., American Society for Microbiology, Washington, DC, 1985, pp. 856-862.

[23] Centers for Disease Control, "Psittacosis Associated with Turkey Processing - Ohio," Morbidity and Mortality Weekly Report, Vol. 30, No. 52, 8 January 1982, pp. 638-640.

[24] Benenson, A. S., Ed., Control of Communicable Diseases in Man, The American Public Health Association, Washington, DC, 1985.

DISCUSSION 1

Does collecting fluid in impingers affect collection/survival of *Chlamydia* and if so what collecting fluid should be used?

CLOSURE 1

Collecting fluid in impingers can affect survival of *C. psittaci*. Normal saline will be detrimental to the collected organisms. I recommend using SPG (sucrose-phosphate-glutamate) medium. The formula for preparing SPG is contained in the manuscript. Rosebury [18] found that viral extract broth was comparable to 10% sheep serum in distilled water for collecting *C. psittaci*. He also showed poor survival in glycerin solutions and in distilled water.

DISCUSSION 2

In regard to airborne *Chlamydia*:

A) is drying of the organisms a problem when sampling with filter membranes;

B) is there a published method of sampling;

C) could you elaborate on interpretation of sampling results; and

D) what is dose/effect or infection threshold, if any?

CLOSURE 2

A) Drying should not be considered a problem when sampling for *C. psittaci* using filter membranes. The organism survives well in the dried state as evidenced by its infectivity via the aerosol route.

B) There has not been a published method for the routine sampling of *C. psittaci* in the indoor environment. Recommendations for monitoring come from aerobiological studies conducted in test chambers, such as those of Rosebury [18].

C) Any evidence of viable *C. psittaci* collected by aerosol sampling is significant. No quantitative evaluation of airborne particles needs to be demonstrated.

D) The human infectious dose of *C. psittaci*, that is, the number of viable organisms necessary to produce serological conversion, has not been accurately determined. It must be remembered that infection does not always result in disease. The development of clinical disease depends to a very large extent on the susceptibility of the host. Thus the highest rates of mortality from psittacosis occur in the elderly whose immune systems are often compromised by aging, immunosuppressive therapy, and/or other chronic and debilitating illnesses.

DISCUSSION 3

Can pet birds carry Chlamydia and be asymptomatic or must they be overtly ill to have the organism?

CLOSURE 3

A major characteristic of virtually all chlamydial infection is latency [11]. Infected pet birds might appear to be healthy, but can develop clinical symptoms and overt disease when stressed. Healthy carriers normally shed fewer chlamydiae than those that are physically ill.

DISCUSSION 4

A) If Chlamydia is transmitted by aerosol, why isn't it a Class III agent?

B) Why do you feel that a dust/mist mask with PF of 10, which is not a HEPA filter material, is sufficient respiratory protection?

C) What is the infectious dose by inhalation of each Chlamydia?

D) What is the minimum particle size collected by the AGI-30 and does efficiency of collection increase with AGIs in series?

E) What is the filter media used in cassette sampling?

F) Elaborate on the PCR analytical technique.

CLOSURE 4

A) Although Chlamydia is transmitted by the aerosol route it normally requires only Biosafety Level 2 practices. These include the use of a Class I or II biological safety cabinet and centrifuge safety precautions. Additional primary containment and personnel precautions, such as those recommended for Biosafety Level 3 may be indicated for activities with high potential for droplet or aerosol production.

B) I do not feel that a dust/mist mask is sufficient protection when monitoring for Chlamydia. I recommend the use of a respirator with HEPA filter cartridges.

C) The human infectious dose or number of viable organisms necessary to produce serological conversion, has not been accurately determined for any of the Chlamydia species.

D) Viruses, smaller than the chlamydiae, have been satisfactorily collected in the AGI-30 prior to assay in tissue culture [Akers, et al, Applied Microbiology, Vol. 14, 1966, pp. 361-364; Akers, et al, Journal of Immunology, Vol. 97, 1966, pp. 379-385]. Some particles, however, of less than 0.3 μm may be carried by the high velocity of the jet air stream through the impinger fluid without being trapped [20]. Efficiency of collection would not necessarily be significantly improved with AGI-30s in series.

E) For cassette sampling, the filters used are made of mixed esters of cellulose.

F) Polymerase chain reaction (PCR) technology is a highly acclaimed technique for rapidly multiplying desired segments of genetic information. It has broad applications in basic research, diagnostics, and forensics. It is anticipated that PCR technology for the detection of _Chlamydia_ will be widely utilized in the future.

DISCUSSION 5

Why operate AGIs for 30 minutes? Doesn't the sampling volume depend on the expected concentration and required sampling sensitivity?

CLOSURE 5

A thirty minute sampling time should be sufficient to collect airborne _Chlamydia_, levels of which could be "high" or "low." Continued sampling beyond 30 minutes might result in the steady decline in viability of collected organisms. Additionally, 30 minutes is the sampling time recommended by the American Conference of Governmental Industrial Hygienists [_Guidelines for the Assessment of Bioaerosols in the Indoor Environment_, ACGIH, Cincinnati, 1989].

DISCUSSION 6

A) Has any study been done to determine how long an aerosol of _Chlamydia_ can survive in the environment, in particular the indoor environment, and still be pathogenic?

B) Would you consider doing air testing in an office setting where conditions seemed appropriate for _Chlamydia_ and if so what criteria should be followed?

CLOSURE 6

A) The survival of aerosolized microorganisms is dependent upon temperature, humidity, airflow, length of time airborne, exposure to ultraviolet radiation, intrinsic resistance to desiccation, and the size of the aerosol cloud. I am not aware of data indicating maximum environmental resistance of airborne _Chlamydia psittaci_ and the ability to cause disease. In addition, the number of _C. psittaci_ required to produce infection (as opposed to disease) in man has not been accurately determined and may vary from strain to strain.

B) Bioaerosol sampling for _C. psittaci_ in an office environment should be considered when the attack rate is significant, symptoms are consistent with psittacosis, and airborne exposure to a bird population has been implicated.

Russell L. Regnery[1] and Joseph E. McDade[2]

Coxiella burnetii (Q FEVER), A POTENTIAL MICROBIAL CONTAMINANT OF THE ENVIRONMENT

REFERENCE: Regnery, R. L. and McDade, J. E., "*Coxiella burnetii* (Q Fever), a Potential Microbial Contaminant of the Environment," *Biological Contaminants In Indoor Environments, ASTM STP 1071*, Philip R. Morey, James C. Feeley, Sr., James A. Otten, Editors, American Society for Testing and Materials, Philadelphia, 1990.

ABSTRACT: Evidence of *Coxiella burnetii* infection is common among certain domestic animals; infected animals can shed large amounts of environmentally stable *C.burnetii*. *C. burnetii* is extremely infectious for humans, and naturally occurring aerosols, as well as other contaminated sources, appear to be responsible for widely dispersed infections of children and adults. Relevant features of what is known about the organism and the human disease it causes are reviewed. Recent findings and areas requiring further investigation are discussed in order to promote a better understanding of the role of *C. burnetii* as a potentially common, infectious contaminant of the environment.

KEYWORDS: *Coxiella burnetii*, Q fever, microbial contaminant, environment, aerosols

Introduction

The human diseases known to be caused by members of the family *Rickettsiaceae* are thought to be transmitted to susceptible vertebrate hosts almost exclusively by specific arthropod vectors. However, in addition to being arthropod-vectored, the agent responsible for Q fever, *Coxiella burnetii*, may also be regarded as a potential contaminant of many environments. The almost worldwide distribution of human Q fever is paralleled by the frequent occurrence of *C. burnetii* infections in several domestic animal species that can serve as reservoirs for dissemination of infectious organisms. Naturally occurring infectious aerosols, originating as a result of excretion of large amounts of

[1]Supervisory Research Microbiologist and [2]Associate Director for Laboratory Science, Center for Infectious Diseases, Centers for Disease Control, Atlanta, GA 30333.

environmentally stable *C. burnetii* during parturition of infected animals, have been documented. The organism is extremely infectious by inhalation. Most infections in humans result in inapparent or self-limiting illness; however, Q fever endocarditis can be life threatening. In this report salient features of Q fever, *C. burnetii*, and the potential for *C. burnetii* transmission as an infectious contaminant of the environment are outlined. A basic understanding of the organism and disease is required before it is possible to interpret the threat of the organism as an environmental contaminant. Information available on infectious *C. burnetii* in the environment is limited, and documentation of human disease is primarily the result of retrospective serologic testing and epidemiologic associations. Reports on environmental sampling methods specifically directed toward *C. burnetii* are few. In this report we stress the need to both further evaluate the role of the disease as a general public health concern and reevaluate the presence of *C. burnetii* in the environment using new, simplified, and sensitive techniques for environmental sampling, particularly those which can also be used to discriminate between *C. burnetii* strains.

Historical Insights

A historical account of the discovery and characterization of Q fever has recently been compiled [1]. Many of the salient features that continue to distinguish Q fever and the causative organism can be understood by reviewing the history of their discovery.

Q fever was first recognized in the mid 1930s in Australia when Edward H. Derrick investigated one of a series of outbreaks of undiagnosed febrile illness among abattoir workers in Brisbane [2]. Derrick began by characterizing the human disease syndrome, and his original descriptions of acute human Q fever are still considered accurate.

Derrick was unable to cultivate bacteria from blood samples from the afflicted slaughterhouse personnel on axenic media. Nor could Derrick demonstrate antibodies to several disease-causing organisms for which tests then existed. However, when he injected either blood or urine from the abattoir workers into guinea pigs, the animals developed a febrile illness [2] which could be experimentally passed to other guinea pigs; in addition, guinea pigs that had recovered were resistant to subsequent infection with the same material. Derrick tentatively referred to the transmissible disease as Query (Q) fever.

Burnet and Freeman, also in Australia, repeated and expanded upon Derrick's experimental results [3]. It is noteworthy that Burnet and Freeman identified, by light microscopy, rickettsia-like organisms within vacuoles of cells from experimentally infected animals. They demonstrated that the infectious agent was able to pass through filters that retained most bacteria, thus giving an indication of the small size of the infectious organism. Immunologic assays showed no cross-reactivity with known rickettsial agents, prompting these investigators to conclude that the new rickettsia-like disease agent was not a member of either the typhus, spotted fever, or scrub typhus sub-groups of rickettsiae [3].

Derrick and coworkers isolated an infectious agent, apparently identical to the Q fever microbe originally isolated from abattoir workers, from ticks removed from native bandicoots, thus suggesting the existence of natural reservoirs of the disease as well as potential arthropod vectors [4]. In addition to infection via the bite of an infected tick, infection by inhalation of aerosolized organisms was also suspected as being an important mode of human acquisition of Q fever infections [5].

Interestingly, the same infectious organism was discovered concurrently in the United States by Davis and Cox [6], although the course of the discovery of Q fever in the United States began not with its appearance as a disease of humans or domestic animals, but rather with an anonymous infectious agent in guinea pigs injected with extracts from a pool of ticks collected in Montana. Ticks were experimentally shown to be capable of vectoring the febrile disease between guinea pigs [7]. Cox and Bell then observed that the organism could be grown in the yolk sacs of embryonated hens' eggs, which provided an important new method for its cultivation [8]. The infectivity for the novel agent for humans became apparent when the then director of the National Institutes of Health, Rolla E. Dyer, acquired a laboratory-associated illness after working with the agent from Montana [9].

The Q Fever Paradigm

The geographic range of Q fever is now recognized as being almost worldwide; the disease is found in at least 51 countries and every continent except Antarctica [10]. Infection of domestic animals, especially cattle, sheep, and goats, as well as of several wild animal species is common although often subclinical [10]. Recovery of infectious organisms from such animals may be difficult; however, typically during and shortly after the animal gives birth, infectious organisms may readily be isolated from several tissue sources including mammary gland (and milk), urine, feces, and especially from the placenta and amniotic fluids [11]. Infectious titers from infected birth products may reach very high levels, exceeding 10^9 infectious particles per gram of tissue [11].

Human infections continue to be reported most frequently in persons closely associated with livestock maintenance and processing [12, 13]. However, retrospective analysis of general populations and hospital admissions demonstrates that Q fever is not limited only to the classic examples of slaughterhouse and laboratory workers [14-17]. Persons not directly associated with livestock are probably infected either by infectious aerosols or by exposure to any of a variety of contaminated secondary sources (e.g., contaminated textiles or clothing [18, 19]) and possibly contaminated raw milk products [20]. Tick transmission of acute Q fever to humans appears to be relatively uncommon despite the fact that ticks have been shown to be effective vectors [7]. *C. burnetii* infections in persons with no obvious contact with *C. burnetii* animal reservoirs or other obvious Q fever risk factors may not be recognized, and hence may be underreported.

Acute Q fever is usually characterized by a variety of signs and symptoms, including sudden onset of fever accompanied by intermittent shaking chills, headache, retrobulbar pain, myalgia, loss of appetite, nausea, joint and chest pains, cough, and often pneumonia or hepatitis [12]. The incubation period before onset of symptoms as well as the duration of the fever are variable, both lasting from a few days to several weeks [12]. Infected persons usually recover fully, although a chronic form of the disease can develop. Cell-mediated immunity appears to play a central role in developing resistance to *C. burnetii* infections [21-23]. However, the role of the immune response during *C. burnetii* infections appears to be complex, and a full understanding of the mechanisms that lead either to naturally acquired immunity or, alternatively, to persistent infections in the presence of high-titer antibody responses is incomplete. Strain differences (see below) and their potentially variable influence on the immune response is another area requiring future investigation.

Chronic Q fever is an apparently uncommon complication of infection, but it is quite serious and frequently fatal. Typically, it occurs in middle-aged or older persons and is often characterized by chronic hepatitis or endocarditis [12]. In such cases, *C. burnetii* organisms may be found in large numbers in heart valves, especially the aortic valve. Antibiotic therapy is less likely to help resolve chronic disease than acute disease. Chronic Q fever patients typically have elevated circulating antibody titers to *C. burnetii* antigens, which suggests that, in these patients, either the antibodies are ineffective in helping to neutralize viable *C. burnetii* or the organisms are inaccessible to the antibodies. Recent data suggest that Q fever endocarditis may be associated with specific *C. burnetii* genotypes (see below).

Q fever as a childhood disease is rather poorly understood. This may be due in part to a historic predilection for identifying the disease within occupation-defined risk groups (e.g., abattoir or farm workers) or the possibility that Q fever among younger persons may produce less obvious disease [24]. However, in the Netherlands, Q fever has been recognized as a disease of infant children requiring hospitalization [15, 25]. In the United States, two separate retrospective analyses of hospital patients from Wisconsin and Pennsylvania both found that children accounted for a large proportion (27% and 55%) of recognized clinical Q fever infections requiring hospitalization [16, 26].

A large serosurvey done in Egypt and Sudan addressed the prevalence of circulating *C. burnetii* antibody in persons of various age groups who lived in close proximity to their livestock [27]. In this study, peak antibody prevalence as determined by the complement fixation (CF) test was found among 2-year olds (47% positive); CF antibody prevalence declined in older age groups, leveling off by the age of 9 years after which prevalence remained relatively constant with age (adult seroprevalence rate approximately 24%). When the same children were bled a second time a few months later, a substantial number, who previously had been seronegative, had seroconverted to *C. burnetii* positive, thus indicating recent infections. These observations suggest that Q fever as a potential childhood disease requires more scrutiny.

Human-to-human transmission is regarded as uncommon, although infected humans can shed viable *C. burnetii* in urine and sputum [24, 28]. Q fever patients and infected cadavers pose a threat of contamination to medical personnel and perhaps to other patients [28].

C. burnetii has been isolated from human placenta and breast milk; however, the possibly serious effects of Q fever on the outcome of human pregnancies has not been completely studied [12].

C. burnetii infections have also been recently reported among immunocompromised adults as well as in an immunocompromised child [29, 30]. Q fever as a possible serious disease of immunocompromised persons within given communities has not yet been evaluated in a systematic manner.

C. burnetii infections are diagnosed on the basis of clinical features coupled with serologic testing (seroconversion is typically regarded as a fourfold rise in specific antibody titer). Serologic techniques include the complement fixation test, immunofluorescence assays, and enzyme-linked immunoassays. None of the tests for circulating antibodies appear to be as sensitive for the detection of prior antigenic stimulus as are assays for cell-mediated immune response [22, 31]. However, because of technical requirements, such cell-mediated immune assays are much more difficult to do than are tests for circulating antibodies; cell mediated immune assays are currently impractical for large scale serosurveillance. In the tests for circulating antibodies, determination of appropriate minimum antibody titers for retrospectively assessing a persons serologic status is clouded by the prospect that, in the quest for increased circulating antibody sensitivity, positive samples probably occur with titers at or below the serum dilution levels that are considered the thresholds of specificity for these testing procedures.

Isolation of infectious *C. burnetii* as a means of diagnosis is rarely attempted because of requirements for cultivation in living cells (e.g., guinea pig, embryonated eggs, or *in vitro* cell culture); this is coupled with concerns regarding biocontainment of amplified organisms within a clinical laboratory. *C. burnetii* grows within vacuoles of infected cells and can be stained with common histologic techniques, Gimenez stain being the stain of choice. However, the small size of the organisms places them at the limits of resolution with the light microscope. Alternatively, *C. burnetii* can be identified in appropriately fixed cells/sections with fluorescein-conjugated specific antibodies.

Like other members of the *Rickettsiaceae* and many other bacteria *C. burnetii* is sensitive to certain antibiotics, including tetracycline, especially during the acute phase of human disease [32-34]. Because of the potential serious consequences of chronic Q fever, antibiotic therapy during acute disease would appear to be a prudent procedure.

Experiments with laboratory maintained-animals and human volunteers have demonstrated that *C. burnetii* is extremely infectious [35, 36]. When injected or inhaled, one infectious particle is capable of initiating infection (as determined by seroconversion). Severity of disease and onset of symptoms may be influenced by dosage and passage history of the organism [35, 36].

Vaccines for Q fever can provide protective immunity in humans [22]. A variety of procedures for vaccine production have been developed, with most of the interest focusing on killed organisms and extracts of inactivated organisms [37]. The vaccines appear to be quite immunogenic and produce a lasting immune response; however, complications of vaccination can occur in persons with prior immunity to the organism. These complications, which occur at the site of vaccination, usually consist of moderately severe local reactions ranging from erythema to sterile abscesses requiring surgical drainage [12]. Such adverse vaccine-associated reactions have in the past compromised the implementation of vaccination programs among high-risk human populations. However, it is now recognized that the sensitivity to adverse vaccination reactions can be predicted by determining a person's preexisting Q fever immune status [38]. When only seronegative abattoir workers, who were also skin-test negative, were vaccinated, no adverse side effects were noted, and no subsequent disease among vaccinees has been reported [38, 39].

Although no commercial vaccine for Q fever is currently licensed in the United States, vaccine is available as part of the Investigative New Drug program of the U.S. Food and Drug Administration. Licensed vaccines for civilian use are currently available in a few other countries such as Australia and Czechoslovakia.

Vaccination of nonhuman, vertebrates that are potential sources of *C. burnetii* has been investigated as a means for decreasing the possibilities of human exposure as well as increasing livestock productivity [40]. As might be expected, animals must be vaccinated prior to acquiring natural infections [41]; if this requirement can be met, it should be possible to develop *C. burnetii*-free animals (e.g., for use in medical research establishments).

C. burnetii Replication, Durability, and Inactivation.

Coxiella burnetii can replicate only within living cells; undoubtedly this has retarded experimental research efforts. Although there are other bacteria that are obligate intracellular parasites--notably members of the genera *Rickettsia*, *Ehrlichia*, and *Wolbachia*--at this time there are no obvious close taxonomic links between *C. burnetii* and these other bacteria (or any free-living bacteria). Recent advances in the genetic analysis of *C. burnetii* may shed new light on the taxonomic/evolutionary status of the organism as specific genes are analyzed and these nucleotide sequences compared with those of other organisms.

The coccoid organism is approximately 0.4-1.0 micron in length and accumulates to large numbers within experimentally infected cells. *C. burnetii* has been reported to undergo an endospore-like differentiation, which may have important implications for the long-term survival of the organism in the environment [42]. *C. burnetii* replicate within phagolysosomes. Recently, it has been demonstrated that the *C. burnetii* organism is metabolically inactive at any pH other than the acid environment associated with the phagolysosome (pH optimum of 4.5) [43]. This ability to be metabolically active at an acid pH appears to be an adaptation for growth within the otherwise hostile phagolysosome environment in which many microbes would be inactivated. Hackstadt and Williams have suggested that this acid pH optimum is a clue to the ability of *C. burnetii* to remain in a metabolically inactive state for long periods of time until the organism once again finds its way into the appropriate cellular habitat; the pH of the phagolysosome may break the otherwise metabolically dormant state of the *C. burnetii* found in the environment and initiate the processes of metabolism and replication [44]. The resistance of *C. burnetii* to chemical degradation within the phagolysosome may also be related to the organism's ability to survive in the environment. The metabolic and chemical stability of infectious *C. burnetii* free in the environment (e.g., as part of an aerosol) underlies the ability of the agent to cause infections far removed from the original source.

The stability of *C. burnetii* in the environment is mirrored by the organism's insensitivity to many commonly used disinfectants [45-47]. Formalin (1% for 24 hours), phenol (1% for 24 hours), alcohol, and quaternary ammonium compounds (1% for 3 hours) are, for example, only marginally effective as *C. burnetii* disinfectants and are not suitable as surface sterilizing agents; chlorine bleach (5.25% hypochlorite), hydrogen peroxide (5%), and Lysol (1%) have been demonstrated in some studies to be reasonably effective as surface disinfectants when applied for reasonably long periods of time (e.g., 3 hours at room temperature) [46, 48, 49]. However, recent experiments may contradict some previous results; 0.5% hypochlorite, 5% Lysol, and 5% formalin were unable to completely inactivate 10^8 infectious *C. burnetii* organisms incubated for 24 hours at 25°C, whereas incubation for 30 minutes at room temperature was sufficient to inactivate the organisms when infectious material was placed in 70% ethyl alcohol, 5% chloroform, or 5% Enviro-chem (a commercial blend of *N*-alkyl dimethyl benzyl and ethylbenzal ammonium chlorides) [47]. In this same study it was found that formaldehyde gas did not sterilize *C. burnetii* in a room without humidity control although *Bacillus* spores were inactivated; humidified formaldehyde gas in an experimental chamber was an effective sterilant (0.3 gram depolymerized paraformaldehyde per cubic foot, maintained overnight). The chemical resistance *C. burnetii* is so extreme that it has been postulated that many of the original "inactivated" *C. burnetii* vaccines (as well as possible lots of smallpox vaccine) contained infectious *C. burnetii* [46].

C. burnetii has been commonly found in samples of commercial raw milk, and raw milk has been implicated as a source of *C. burnetii* infection [20, 50, 51]. However, in two studies in which volunteers were fed milk known to contain infectious *C. burnetii*, no disease was noted [52, 53]. No hypothesis has been tested to account for this lack of experimentally induced disease; perhaps induction of Q fever *per os* is either dose dependent and/or strain dependent. Alternatively, it is possible that Q fever associated with milk products is actually the result of inadvertent aerosolization and inhalation of infections organisms. Treatment of milk at 63°C for 30 minutes or 71.6°C for 15 seconds is required to inactivate *C. burnetii* [54].

Genetic Analysis

Until recently it was thought that *C. burnetii* was an entirely monotypic species, devoid of any strain-specific markers. Recently, however, it has been demonstrated that different isolates of *C. burnetii* may indeed have subtle genotypic and phenotypic differences. Mallavia and coworkers have demonstrated that DNA of purified *C. burnetii* from different sources may yield slightly different patterns when digested with restriction endonucleases [55]. These same investigators discovered plasmids (extrachromosomal circular DNA) in isolates of *C. burnetii* [56]. Importantly, these genotypic differences may correlate to some extent with their vertebrate hosts of origin, as well as the severity of disease [57]. At least three different genotypes have been recognized by genome DNA restriction endonuclease digest patterns and plasmid characterization [57]. All isolates from human endocarditis cases fall into one of two groups, as defined by plasmid type and by genome restriction endonuclease fragment patterns. Plasmids have not been identified in any other rickettsia-like organism, once again emphasizing the differences between *C. burnetii* and other organisms with similar ecologies.

Recently, Moos and Hackstadt have reported that subtle differences exist between the lipopolysaccharide molecules of certain *C. burnetii* isolates [58]. All *C. burnetii* isolates from human endocarditis patients have a unique lipopolysaccharide. The *C. burnetii* isolates from endocarditis cases, which fit into the unique lipopolysaccharide phenotype cluster, correlate exactly with the unique plasmid, genotype-defined endocarditis isolates, thus reinforcing the significance of both sets of observations.

Q Fever: Recent Examples

Outbreaks of Q fever continue despite our increased understanding of the disease. Abattoir workers and those persons working directly with parturient livestock still remain among the highest risk groups [13, 59]. However, the disease is not limited to these "historically significant" Q fever populations. A few selected recent reports illustrate additional examples of continued potential public health problems associated with Q fever.

Sheep in Switzerland are often brought down from alpine pastures in the fall of the year to lowland grazing areas. During the 1983 autumn movement of sheep, approximately 850 sheep were herded through a sub-alpine Swiss valley community [60, 61]. Within 3 weeks of the passage of the sheep, 191 acute Q fever cases occurred among villagers (total population 4652; 4.1% acute infection rate). Although this study has been criticized for incomplete statistical analysis [62], the temporal association between sheep transit and acute Q fever in the community remains convincing. This outbreak is one relatively recent example of the ability of *C. burnetii* to infect persons on a community level; the organisms are thought to have been disseminated as widespread infectious aerosols. Other recent reports suggest that community-wide outbreaks of Q fever have occurred within relatively urban settings of the United Kingdom; no common source of infection has yet been specifically identified [63, 64].

In 1980, an outbreak of Q fever occurred in a medical school in Denver in which pregnant sheep were used in medical research experiments [65]. Sheep were being transported through the building's elevator, and 137 persons were infected (as defined by rising antibody titers), resulting in 65 recognized clinical cases. Only 41 of these persons had direct contact with the sheep. A similar outbreak occurred in a San Francisco medical school during 1979 [66]. In this example, 144 persons had serologic evidence of recent Q fever infections and there were 88 clinical cases. Again, only a fraction of the persons had any direct sheep contact. In both of these medical school outbreaks, epidemiologic evidence suggests that the organisms were spread via infectious aerosols through the buildings' air-handling systems.

The potential for just such large-scale indoor Q fever outbreaks had been previously addressed in a publication [67]. Subsequent to the Denver outbreak, specific recommendations for the use of sheep in research establishments were formulated [68]. These recommendations include 1) maintaining physical separation between sheep research/animal care facilities and other laboratories or patient areas, 2) designation of such research areas as biohazard areas for Q fever, and 3) appropriate decontamination of potentially infected material (e.g., contaminated laundry, surgical instruments, animal tissue). If physical separation between facilities cannot be maintained, additional precautions are recommended; these include 1) transporting of sheep only through areas designated as Q fever biohazard areas, 2) controlled exhaust-air ventilation systems that preclude circulation of infectious aerosols to other areas of a facility, and 3) maintenance of negative air pressures in Q fever biohazard areas relative to non-biohazard areas. There are also multiple procedural recommendations that deal with disinfection/decontamination, techniques with working with biohazardous materials, and notification of risk potential to personnel (including women of childbearing age who have a theoretical risk of fetal infection during pregnancy). Despite the recognition of the risks of Q fever, new research-associated outbreaks continue to be reported [69].

Domestic cats have recently been implicated as a primary source of human *C. burnetii* infections in Nova Scotia [70]. For example, in one outbreak, 16 of 32 persons employed at a large truck repair facility became infected as a result of secondary, passive *C. burnetii* distribution via the clothing of the owner of a parturient cat [19]. Another Nova Scotia Q fever outbreak occurred among 12 friends who played poker together in the same room where a *C. burnetii*-infected cat was delivering kittens [71]. Currently, the distribution of *C. burnetii* among cat populations, other than in Nova Scotia, has not been reported. The possible implications for cat-associated Q fever are significant if additional cat populations are found to be capable of sustaining *C. burnetii* infections and transmitting the disease to humans.

These few examples illustrate the variety of situations in which multiple, simultaneous Q fever infections have occurred and in which infections probably did not occur as the result of direct contact with copious amounts of infected animal tissue but rather as a result of *C. burnetii* contamination of secondary sources or production of infectious aerosols. It is difficult to assess the prevalence of scattered, individual human infection and disease; however, there seems to be no reason to suspect that the disease is limited to multiple-infection outbreaks. Perhaps one of the greatest outstanding questions regarding the epidemiology of Q fever is not how infection can occur, but rather, why are there not more Q fever cases recognized?

Attempts at Estimating the Prevalence of *C. burnetii*

One means of estimating how common *C. burnetii* may be in the environment would be to determine how common are *C. burnetii* infections of the domestic animals to which people are most often exposed. As one might expect, the incidence of seropositive domestic animals varies between animal types, location, testing procedures, and probably many other factors. Clearly, however, serologic evidence of Q fever infections in domestic animals is not uncommon. In one study, for example, between 32% and 73% of Wisconsin dairy cattle were shown to be *C. burnetii* seropositive, and 84% of the cows from seropositive herds were actively shedding *C. burnetii* in their milk [50]. In a California study, 82% of dairy cows tested were positive for *C. burnetii* antibody, and 23% of the animals were actively shedding *C. burnetii* in their milk [51]. Another study from California concluded that 99% of pooled specimens from various dairy herds were seropositive [17]. In a Nova Scotia serosurvey, 6.7% of sheep, 7% of goats, 24% of cattle, and 24.1% of cats were positive for *C. burnetii* circulating antibodies [72].

It is difficult to monitor *C. burnetii* infections in non-occupationally defined human populations with presumably no obvious source of exposure and infection. As noted previously, Q fever infections often resolve themselves uneventfully without medical intervention. Limited data exist concerning general seroprevalence of *C. burnetii* antibodies. Examples of such data from different locations suggest that human seroprevalence occurs at significant levels (e.g., between 2% and 30%) among general populations in areas where *C. burnetii* is endemic [15-17, 27, 73].

As noted above, the several techniques used to detect significant antibody titers vary in sensitivity and specificity. It is not the purpose of this report to review either the methods used for detection of past infection with *C. burnetii* or the important criteria used to establish what is regarded as minimum positive results for each test. However, it should be mentioned that, although the complement fixation test is very specific, it is relatively insensitive for detection of *C. burnetii* antibodies compared to fluorescent antibody (FA) techniques [60] and neither test appears to be as sensitive as assays that detect cell-mediated immune response [21-23, 74]. Hence many of the older reports of seropositivity based on complement fixation tests may be considered underestimates when compared to with data from FA tests or other more sensitive assays. For example, in one study of dairymen in the Milwaukee area, 14% had demonstrable antibodies by the complement fixation test; however, when the same men were skin tested for *C. burnetii*, 98% were positive [31].

Similarly, in a recent vaccine study, whereas only 50%-60% of vaccinees developed complement-fixing or immunofluorescent antibodies to *C. burnetii* after vaccination, 85%-95% of these apparent non-responders exhibited *in vitro* cell-mediated immunologic responses to *C. burnetii* antigen [22]. Importantly, statistical evidence suggests that these persons with cell-mediated responses, but no apparent circulating antibodies, were protected from naturally occurring infections when compared with an unvaccinated control group. However, although assays for cell-mediated immune responses seem to be the most sensitive techniques for detection of prior exposure to *C. burnetii* antigens, these techniques are often impractical to conduct on a large scale.

Environmental Monitoring

Relatively little has been published regarding the monitoring of *C. burnetii* in the environment. The original reports demonstrating naturally occurring aerosols and the presence of *C. burnetii* in soil and water samples remain the only published information available. In all of these original environmental sampling studies, the presence of the *C. burnetii* organisms was detected by infecting either guinea pigs or hamsters with sample aliquots and subsequently monitoring the animal for the presence of diagnostic antibody production. Methods previously used for collecting aerosolized *C. burnetii* specimens are standard techniques for monitoring airborne microbes. Because of the absence of reports comparing aerosol sampling techniques for *C. burnetii*, we believe it would be inappropriate at this time to advocate testing standards, especially in lieu of the advent of newer sampling methods that have not been applied to *C. burnetii* sampling. A brief summary of the available data is given in the paragraphs below.

Dust samples containing infectious *C. burnetii* have been collected on the premises of a northern California sheep ranch as well as a southern California dairy [75]. A similar study demonstrated infectious *C. burnetii* aerosols in a barn housing goats [76]. Capillary impinger air-sampling apparatus were used to collect these aerosol samples into liquid medium prior to their injection into animals for detection of infectious organisms.

In a controlled experimental study, cotton air filters were used to trap infectious *C. burnetii* aerosols that persisted in sheep stalls for up to 2 weeks after infected ewes gave birth [77]. The cotton filters were washed and the eluvium used to infect guinea pigs.

C. burnetii was isolated in hamsters inoculated with water samples from 6 of 19 surface water pools found on sheep ranches in California [78]. It was determined that infectious *C. burnetii* titers ranged from 30 to 10,000 infectious hamster doses per milliliter of water sample. In the same study it was shown that soil samples from the sheep ranches were likewise infectious and that recovery of *C. burnetii* from soil samples paralleled the lambing activity on the ranches. Infectious titers were not determined for soil samples.

Guinea pigs and humans are reported to be equally susceptible to infection by inhaled *C. burnetii* [35]. These observations suggest that it is theoretically possible to use guinea pigs as sentinels for infectious aerosols. Such a sentinel animal system would, like previous *C. burnetii* environmental monitoring efforts, be based on serologic monitoring of the guinea pigs. It is noteworthy that the use of sentinel animals is considered a standard technique for monitoring and detection of arthropod-borne viruses in the environment [79].

The small size of the *C. burnetii* organism is important from the standpoint that individual organisms are at the limits of resolution by light microscopy. Direct microscopic detection and recognition of *C. burnetii* organisms as part of environmental samples collected on filters (and possibly stained with specific fluorescent antibodies), although theoretically possible, would be difficult because of the small size and lack of definitive morphology of isolated organisms. Identification of fluorescent-stained *C. burnetii* fixed within infected host cells, with multiple organisms clustered together inside cytoplasmic vacuoles, is considerably different from attempting to make a definitive identification based on an isolated organism(s) collected on a filter along with additional environmental detritus. However, particle counts of fluorescent stained *C. burnetii* retained on Nucleopore filters have been used for specialized, quantitative experimental purposes [58].

Any discussion of techniques for detecting *C. burnetii* must include some mention of the risk of Q fever to laboratory workers; such workers constitute one of the highest Q fever risk groups [12]. Experiments with *C. burnetii* should be performed only in biocontainment level-3 (BCL-3) laboratories [80]. Such practical requirements stipulate limited personnel access, negative air-flow, and HEPA-filtered biocontainment biological safety cabinets to ensure relatively safe handling of infectious organisms. In addition, personnel should use back-fastening gowns, masks, and, most importantly, careful technique. Animal experiments conducted within such biocontainment facilities are expensive and become quite involved logistically. Unlike many other microbes treated in BCL-3 facilities, *C. burnetii* is resistant to disinfection (as noted above), which further complicates laboratory safety procedures. These considerations limit the number of modern laboratories that engage in *C. burnetii* research.

The recently developed polymerase chain reaction (PCR) technique to detect minute quantities of genetic material is currently being evaluated as a method for detecting the presence of *C. burnetii* in clinical samples; presumably the same or similar techniques could be used for monitoring potential environmental sources of infection [81]. Large numbers of samples, collected by any one of a variety of techniques, could be rapidly evaluated; the PCR technique will undoubtedly be even further automated in the near future. In addition to the relative ease with which samples can be tested using PCR technology, this technique minimizes the need to work with large amounts of infectious organisms (specimens could be inactivated prior to testing). The PCR and related techniques may well usher in a new recognition and appreciation of the prevalence of *C. burnetii* as an environmental contaminant.

Furthermore, it has been demonstrated that in similar microbial systems, PCR-amplified DNA products can be used as substrata for restriction endonuclease cleavage and diagnostic evaluation of species and genotypes [82]. This would be especially appropriate for the study of *C. burnetii* in the environment since, as noted above, genotypic variations may correlate with the potential for serious disease in humans [57, 83].

Summary

Several lines of evidence suggest that *C. burnetii* can be regarded as a rather common environmental contaminant. Infections caused by this organism appear to be relatively common in species of domestic animals and can pose a threat to humans, especially when such domestic animals are in close proximity to large human populations. Large amounts of environmentally stable, highly infectious organisms can be shed by animal hosts during parturition. Limited data indicate that, in addition to the recognized classic high-risk groups, many persons (including children), are exposed to at least low levels of infectious organisms. What is recognized as either acute or chronic Q fever apparently accounts for a minority of actual human infections. A more complete understanding of human Q fever and the role of *C. burnetii* as a contaminant of the environment depends on several advances in Q fever research. These steps will include 1) a better understanding of immune host responses to low doses of inhaled *C. burnetii*, 2) enhanced, standardized immunologic means to reliably detect prior exposure, 3) a better understanding of variation in pathogenicity between strains, and 4) better systems for detecting *C. burnetii* in the environment. Until these conditions are satisfied, the true role of *C. burnetii* in the environment as a potential source of human infection will continue to be incompletely understood and appreciated.

REFERENCES

[1] McDade, J. E., "Historical aspects of Q fever," in *Q Fever, Volume I: The Disease*, Marrie, T. J. (ed.), CRC Press, Inc., Boca Raton, 1990.

[2] Derrick, E. H., "'Q' fever, a new fever entity: clinical features, diagnosis and laboratory investigation," *Medical Journal of Australia*, Vol. 2, 1937, pp. 281-299.

[3] Burnet, F. M. and Freeman, M., "Experimental studies on the virus of "Q" fever," *Medical Journal of Australia*, Vol. 2, 1937, pp. 299-305.

[4] Derrick, E. H. and Smith, D. J. W., "Studies in the epidemiology of Q fever II: The isolation of three strains of *Rickettsia burneti* from the bandicoot *Isoodon torosus*," *Australian Journal of Experimental Biology and Medical Science*, Vol. 18, 1940, pp. 99-102.

[5] Burnet, F. M. and Freeman, M., "Note on a series of laboratory infections with the rickettsia of Q fever," *Medical Journal of Australia*, Vol. 1, 1939, pp. 11-12.

[6] Davis, G. E. and Cox, H. R., "A filter-passing infectious agent isolated from ticks I: Isolation from *Dermacentor andersoni*, reactions in animals, and filtration experiments," *Public Health Reports*, Vol. 53, 1938, pp. 2259-2267.

[7] Parker, R. R. and Davis, G. E., "A filter-passing infectious agent isolate from ticks II: transmission by *Dermacentor andersoni*," *Public Health Reports*, Vol. 53, 1938, pp. 2267-2270.

[8] Cox, H. R. and Bell, E. J., "The cultivation of *Rickettsia diaporica* in tissue culture and in the tissues of developing chick embryos," *Public Health Reports*, Vol. 54, 1939, pp. 2171-2178.

[9] Dyer, R. E., "A filter-passing infectious agent isolated from ticks IV: human infection," *Public Health Reports*, Vol. 53, 1938, pp. 2277-2282.

[10] Kaplan, M. M. and Bertagna, P., "The geographical distribution of Q fever," *Bulletin of the World Health Organization*, Vol. 13, 1955, pp. 829-860.

[11] Welsh, H. H., Lennette, E. H., Abinanti, F. R., and Winn, J. F., "Q fever in California IV: Occurrence of *Coxiella burnetii* in the placenta of naturally infected sheep," *Public Health Reports*, Vol. 66, 1951, pp. 1473-1477.

[12] Sawyer, L. A., Fishbein, D. B., and McDade, J. E., "Q fever: current concepts," *Reviews of Infectious Diseases*, Vol. 9, 1987, pp. 935-946.

[13] Centers for Disease Control, "Q fever among slaughterhouse workers--California," *Morbidity and Mortality Weekly Reports.*, Vol. 35, 1986, pp. 223-226.

[14] Beck, D. M., Bell, J. A., Shaw, E. W., and Huebner, R. J., "Q fever studies in Southern California II: an epidemiological study of 300 cases," *Public Health Reports*, Vol. 64, 1949, pp. 41-56.

[15] Richardus, J. H., Donkers, A., Dumas, A. M., Schaap, G. J., Akkermans, J. P., Huisman, J., and Valkenburg, H. A., "Q fever in the Netherlands: a sero-epidemiological survey among human population groups from 1968 to 1983," *Epidemiology and Infection*, Vol. 98, 1987, pp. 211-219.

[16] Wisniewski, H. J. and Krumbiegel, E. R., "Q fever in the Milwaukee area III: Epidemiological studies of Q fever in humans," *Archives of Environmental Health*, Vol. 21, 1970, pp. 66-70.

[17] Gross, P. A., Portnoy, B., Salvatore, M. A., Kogan, B. A., Heidbreder, G. A., Schroeder, R. J., and McIntyre, R. W., "Q fever in Los Angeles County: serological survey of human and bovine populations," *California Medicine: The Western Journal of Medicine*, Vol. 114, 1971, pp. 12-15.

[18] Sigel, M. M., Scott, T. F. M., and Henle, W., "Q fever in a wool and hair processing plant," *American Journal of Public Health*, Vol. 40, 1950, pp. 524-532.

[19] Marrie, T. J., Langille, D., Papukna, V., and Yates, L., "Truckin' pneumonia--an outbreak of Q fever in a truck repair plant probably due to aerosols from clothing contaminated by contact with newborn kittens," *Epidemiology and Infection*, Vol. 102, 1989, pp. 119-127.

[20] Brown, G. L., Colwell, D. C., and Hooper, W. L., "An outbreak of Q fever in Staffordshire," *Journal of Hygiene*, Vol. 66, 1968, pp. 649-655.

[21] Jerrells, T. R., Mallavia, L. P., and Hinrichs, D. J., "Detection of long-term cellular immunity to *Coxiella burnetii* as assayed by lymphocyte transformation," *Infection and Immunity*, Vol. 11, 1975, pp. 280-286.

[22] Izzo, A. A., Marmion, B. P., and Worswick, D. A., "Markers of cell-mediated immunity after vaccination with an inactivated, whole-cell Q fever vaccine," *Journal of Infectious Diseases*, Vol. 157, 1988, pp. 781-789.

[23] Kazar, J., "Immunity in Q fever. Czechoslovakia," *Acta Virologica*, Vol. 32, 1988, pp. 358-368.

[24] Derrick, E. H., "The course of infection with *Coxiella burneti*," *The Medical Journal of Australia*, Vol. 1, 1973, pp. 1051-1057.

[25] Richardus, J. H., Dumas, A. M., Huisman, J., and Schaap, G. J., "Q fever in infancy: a review of 18 cases," *Pediatric Infectious Disease*, Vol. 4, 1985, pp. 369-373.

[26] Marshak, R. R., Melbin, J., and Herman, M. J., "Study of Q fever in animals and man in Pennsylvania," *American Journal of Public Health*, Vol. 51, 1961, pp. 1189-1198.

[27] Taylor, R. M., Kingston, J. R., and Rizk, F., "Serological (complement-fixation) surveys for Q fever in Egypt and the Sudan, with special reference to its epidemiology in areas of high endemicity," *Archives de l'Institut Pasteur de Tunis.*, Vol. 36, 1959, pp. 529-556.

[28] Derrick, E. H., "The epidemiology of "Q" fever: a review," *The Medical Journal of Australia*, Vol. 1, 1953, pp. 245-253.

[29] Loudon, M. M. and Thompson, E. N., "Severe combined immunodeficiency syndrome, tissue transplant, leukaemia, and Q fever," *Archives of the Diseases of Children*, Vol. 63, 1988, pp. 207-209.

[30] Heard, S. R., Ronalds, C. J., and Heath, R. B., "*Coxiella burnetii* infection in immunocompromised patients," *Journal of Infection*, Vol. 11, 1985, pp. 15-18.

[31] Wisniewski, H. J., Kleiman, M. M., Lackman, D. B., and Krumbiegel, E. R., "Demonstration of inapparent infection with disease agents common to animals and man," *Health Laboratory Sciences*, Vol. 6, 1969, pp. 173-177.

[32] Spicer, A. J., Peacock, M. G., and Williams, J. C., "Effectiveness of several antibiotics in suppressing chick embryo lethality during experimental infections by *Coxiella burnetii*, *Rickettsia typhi*, and *R. rickettsii*," in *Rickettsiae and rickettsial diseases*, Burgdorfer, W. and Anacker, R. L. (ed.). Academic Press, New York, 1981, pp. 375-383.

[33] Yeaman, M. R., Roman, M. J., and Baca, O. G., "Antibiotic susceptibilities of two *Coxiella burnetii* isolates implicated in distinct clinical syndromes," *Antimicrobial Agents and Chemotherapy*, Vol. 33, 1989, pp. 1052-1057.

[34] Raoult, D., "Antibiotic susceptibility of rickettsia and treatment of rickettsioses," *European Journal of Epidemiology*, Vol. 5, 1989, pp. 432-435.

[35] Tigertt, W. D., Benenson, A. S., and Gochenour, W. S., "Airborne Q fever," *Bacteriological Reviews*, Vol. 25, 1961, pp. 285-293.

[36] Ormsbee, R., Peacock, M., Gerloff, R., Tallent, G., and Wike, D., "Limits of rickettsial infectivity," *Infection and Immunity*, Vol. 19, 1978, pp. 239-245.

[37] Marmion, B. P., "Development of Q-fever vaccines, 1937 to 1967," *The Medical Journal of Australia*, Vol. 2, 1984, pp. 1074-1078.

[38] Marmion, B. P., Ormsbee, R. A., Kyrkou, M., Wright, J., Worswick, D., Cameron, S., Esterman, A., Feery, B., and Collins, W., "Vaccine prophylaxis of abattoir-associated Q fever," *Lancet*, Vol. 2, 1984, pp. 1411-1414.

[39] Kazar, J., Schramek, S., and Brezina, R., "The value of skin test in Q fever convalescents and vaccinees as indicator of antigen exposure and inducer of antibody recall," *Acta Virologica*, Vol. 28, 1984, pp. 134-140.

[40] Aitken, I. D., "Clinical aspects and prevention of Q fever in animals," *European Journal of Epidemiology*, Vol. 5, 1989, pp. 420-424.

[41] Schmeer, N., Muller, P., Langel, J., Krauss, H., Frost, J. W., and Wieda, J., "Q fever vaccines for animals," *Zentralblatt fur Bakteriol. Mikrobiol. Hyg.*, Vol. 267, 1987, pp. 79-88.

[42] McCaul, T. F. and Williams, J. C., "Developmental cycle of *Coxiella burnetii*--structure and morphogenesis of vegetative and sporogenic differentiations," *Journal of Bacteriology*, Vol. 147, 1981, pp. 1063-1076.

[43] Hackstadt, T. and Williams, J. C., "pH dependence of the *Coxiella burnetii* glutamate transport system," *Journal of Bacteriology*, Vol. 154, 1983, pp. 598-603.

[44] Hackstadt, T. and Williams, J. C., "Biochemical stratagem for obligate parasitism of eukaryotic cells by *Coxiella burnetii*," *Proceedings of the National Academy of Sciences of the United States of America*, Vol. 78, 1981, pp. 3240-3244.

[45] Ransom, S. E. and Huebner, R. J., "Studies on the resistance of *Coxiella burneti* to physical and chemical agents," *American Journal of Hygiene*, Vol. 53, 1951, pp. 110-119.

[46] Malloch, R. A. and Stoker, M. G. P., "Studies on the susceptibility of *Rickettsia burneti* to chemical disinfectants, and on techniques for detecting small numbers of viable organisms," *Journal of Hygiene*, Vol. 50, 1952, pp. 502-514.

[47] Scott, G. H. and Williams, J. C., "Susceptibility of *Coxiella burnetii* to chemical disinfectants", *Annals of the New York Academy of Sciences, Rickettsiology: Current Issues and Perspectives, Proceedings of the Eighth Sesqui-annual Meeting of the American Society for Rickettsiology and Rickettsial Diseases*, Hechemy, K. E., Paretsky, D, Walker, D. H., and Mallavia, L. P. (ed.), Vol. 590, 1989, pp. 291-296.

[48] Babudieri, B., "Q fever: A zoonosis," *Advances in Veterinary Science*, Vol. 5, 1959, pp. 82-182.

[49] Ormsbee, R. A., "Q fever rickettsia," in *Viral and Rickettsial Infections of Man*, Horsfall, F. and Tamm, I., (eds), edition 4, JB Lippincott Company, Philadelphia, 1965, pp. 1144-1160.

[50] Wisniewski, H. J. and Krumbiegel, E. R., "Q fever in the Milwaukee area I: Q fever in Milwaukee area cattle," *Archives of Environmental Health*, Vol. 21, 1970, pp. 58-62.

[51] Biberstein, E. L., Behymer, D. E., Bushnell, R., Crenshaw, G., Riemann, H. P., and Franti, C. E., "A survey of Q fever (*Coxiella burnetii*) in California dairy cows," *American Journal of Veterinary Research*, Vol. 35, 1974, pp. 1577-1582.

[52] Benson, W. W., Brock, D. W., and Mather, J., "Serologic analysis of a penitentiary group using raw milk from a Q fever infected herd," *Public Health Reports*, Vol. 78, 1963, pp. 707-710.

[53] Krumbiegel, E. R. and Wisniewski, H. J., "Q fever in Milwaukee II: Consumption of infected raw milk by human volunteers," *Archives of Environmental Health*, Vol. 21, 1970, pp. 63-65.

[54] Enright, J. B., Sadler, W. W., and Thomas, R. C., "Pasteurization of milk containing the organism of Q fever," *American Journal of Public Health*, Vol. 47, 1957, pp. 695-700.

[55] Mallavia, L. P., Samuel, J. E., and Frazier, M. E., "Conservation of plasmid DNA sequences in various Q fever isolates," in *Rickettsiae and Rickettsial Diseases: Proceedings of the Third International Symposium* Kazar, J. (ed.), Slovak Academy of Sciences, Bratislava, 1985, pp. 137-145.

[56] Samuel, J. E., Frazier, M. E., Kahn, M. L., Thomashow, L. S., and Mallavia, L. P., "Isolation and characterization of a plasmid from phase I *Coxiella burnetii*," *Infection and Immunity*, Vol. 41, 1983, pp. 488-493.

[57] Samuel, J. E., Frazier, M. E., and Mallavia, L. P., "Correlation of plasmid type and disease caused by *Coxiella burnetii*," *Infection and Immunity*, Vol. 49, 1985, pp. 775-779.

[58] Moos, A. and Hackstadt, T., "Comparative virulence of intra- and interstrain lipopolysaccharide variants of *Coxiella burnetii* in the guinea pig model," *Infection and Immunity*, Vol. 55, 1987, pp. 1144-1150.

[59] Rauch, A. M., Tanner, M., Pacer, R. E., Barrett, M. J., Brokopp, C. D., and Schonberger, L. B., "Sheep-associated outbreak of Q fever, Idaho," *Archives of Internal Medicine*, Vol. 147, 1987, pp. 341-344.

[60] Dupuis, G., Peter, O., Pedroni, D., and Petite, J., "Clinical aspects observed during an epidemic of 415 cases of Q fever," *Schweiz. Med. Wochenschr.*, Vol. 115, 1985, pp. 814-818.

[61] Centers for Disease Control, "Q fever outbreak--Switzerland," *Morbidity and Mortality Weekly Reports*, Vol. 33, 1984, pp. 355-356,361.

[62] Senn, S., "A note concerning the analysis of an epidemic of Q fever," *International Journal of Epidemiology*, Vol. 17, 1988, pp. 891-893.

[63] Smith, G., "Q fever outbreak in Birmingham, UK [letter]," *Lancet*, Vol. 2, 1989, p. 557.

[64] Salmon, M. M., Howells, B., Glencross, E. J., Evans, A. D., and Palmer, S. R., "Q fever in an urban area," *Lancet*, Vol. 1, 1982, pp. 1002-1004.

[65] Meiklejohn, G., Reimer, L. G., Graves, P. S., and Helmick, C., "Cryptic epidemic of Q fever in a medical school," *Journal of Infectious Diseases*, Vol. 144, 1981, pp. 107-113.

[66] Spinelli, J. S., Ascher, M. S., Brooks, D. L., Dritz, S. K., Lewis, H. A., Morrish, R. H., Rose, L., and Ruppanner, R., "Q fever crisis in San Francisco--Controlling a sheep zoonosis in a lab animal facility," *Lab Animal*, Vol. 15, 1981, pp. 24-27.

[67] Schachter, J., Sung, M., and Meyer, K. F., "Potential danger of Q fever in a university hospital environment," *Journal of Infectious Diseases*, Vol. 123, 1971, pp. 301-304.

[68] Bernard, K. W., Parham, G. L., Winkler, W. G., and Helmick, C. G., "Q fever control measures: recommendations for research facilities using sheep," *Infection Control*, Vol. 3, 1982, pp. 461-465.

[69] Graham, C. J., Yamauchi, T., and Rountree, P., "Q fever in animal laboratory workers: an outbreak and its investigation," *American Journal of Infection Control*, Vol. 17, 1989, pp. 345-348.

[70] Marrie, T. J., Durant, H., Williams, J. C., Mintz, E., and Waag, D. M., "Exposure to parturient cats: a risk factor for acquisition of Q fever in Maritime Canada," *Journal of Infectious Diseases*, Vol. 158, 1988, pp. 101-108.

[71] Langley, J. M., Marrie, T. J., Covert, A., Waag, D. M., and Williams, J. C., "Poker players' pneumonia. An urban outbreak of Q fever following exposure to a parturient cat," *New England Journal of Medicine*, Vol. 319, 1988, pp. 354-356.

[72] Marrie, T. J., Van Buren, J., Fraser, J., Haldane, E. V., Faulkner, R. S., Williams, J. C., and Kwan, C., "Seroepidemiology of Q fever among domestic animals in Nova Scotia," *American Journal of Public Health*, Vol. 75, 1985, pp. 763-766.

[73] Marrie, T. J., "Seroepidemiology of Q fever in New Brunswick and Manitoba," *Canadian Journal of Microbiology*, Vol. 34, 1988, pp. 1043-1045.

[74] Bell, F. J., Luoto, L., Casey, M., and Lackman, D. B., "Serologic and skin-test response after Q fever vaccination by the intracutaneous route," *Journal of Immunology*, Vol. 93, 1964, pp. 403-408.

[75] DeLay, P. D., Lennette, E. H., and DeOme, K. B., "Q fever in California II: recovery of *Coxiella burneti* from naturally-infected air-borne dust," *Journal of Immunology*, Vol. 65, 1950, pp. 211-220.

[76] Lennette, E. H. and Welsh, H. H., "Q fever in California X: Recovery of *Coxiella burneti* from the air of premises harboring infected goats," *American Journal of Hygiene*, Vol. 54, 1951, pp. 44-49.

[77] Welsh, H. H., Lennette, E. H., Abinanti, F. R., and Winn, J. F., "Air-borne transmission of Q fever: the role of parturition in the generation of infective aerosols," *Annals of the New York Academy of Sciences*, Vol. 70-71, 1958, pp. 528-540.

[78] Welsh, H. H., Lennette, E. H., Abinanti, F. R., Winn, J. F., and Kaplan, W., "Q fever studies XXI: the recovery of *Coxiella burnetii* from the soil and surface water of premises harboring infected sheep," *American Journal of Hygiene*, Vol. 70, 1959, pp. 14-20.

[79] Sudia, W. D., Lord, R. D., and Hayes, R. O., *Collection and Processing of Vertebrate Specimens for Arbovirus Studies*, U.S. Department of Health, Education, and Welfare; Public Health Service, Atlanta, 1972.

[80] Richardson, J. H. and Barkley, W. E., *Biosafety In Microbiological and Biomedical Laboratories*, U.S. Government Printing Office, Washington, D.C., 1984.

[81] Frazier, M. E., Mallavia, L. P., Samuel, J. E., and Baca, O. G., "DNA probes for identification of *Coxiella burnetii* strains," *Annals of the New York Academy of Sciences, Rickettsiology: Current Issues and Perspectives, Proceedings of the Eighth Sesqui-annual Meeting of the American Society for Rickettsiology and Rickettsial Diseases*, Hechemy, K. E., Paretsky, D, Walker, D. H., and Malavia, L. P. (ed.), Vol. 590, 1989, pp. 445-458.

[82] Regnery, R,. L., Spruill, C. L., and Plikaytis, B. D., "Genotypic identification of rickettsiae and estimation of intraspecies sequence divergence for portions of two rickettsial genes," in preparation.

[83] Vodkin, M. H., Williams, J. C., and Stephenson, E. H., "Genetic heterogeneity among isolates of *Coxiella burnetii*," *Journal of General Microbiology*, Vol. 132, 1986, pp. 455-463

DISCUSSION

Discussion: After having once been infected, are people no longer susceptible to reinfection with *C. burnetii*?

Closure: Appropriate vaccination of abattoir workers appears to produce long term immunity [38] and presumably long term immunity likewise results from naturally occurring infections from which persons fully recover. True reinfection of completely recovered individuals (as contrasted with latent or sequestered infections) has not been documented. The possibility of recurring antigenic stimulation of a previously immunized individual by small doses of an organism, that may be considered an environmental contaminant under appropriate conditions, may possibly contribute to the maintenance of long term immunity.

Discussion: Since most Q fever infections are not life-threatening and, like other rickettsial diseases, can be treated successfully with antibiotics (at least early during the course of infection) is it important to sample and identify *C. burnetii* to protect the health of the public?

Closure: The answer to this question depends considerably upon one's perspective. A good case can be made that *C. burnetii* is a rather common contaminant of certain environments. As currently recognized, acute Q fever (excluding more serious forms of disease) can produce transient, serious debilitating infections (explaining in part the past interest of *C. burnetii* as an agent of biological warfare). Mortality is not the only standard by which a disease can be measured. At this point in time, the significance of possible sub-acute infections to the overall health of individuals has not been evaluated even from the standpoint of documenting whether such infections are truly inapparent or perhaps mistaken for other common infections (e.g., 'flu-like disease). Under conditions where outbreaks of Q fever are known to have occurred or where they could easily occur in the future, prudence suggests taking steps to preclude future disease episodes (such as those guidelines formulated for those doing sheep research, [68]). However, the role of Q fever as an infection of the general public, and the steps that should be taken to better understand and possibly control the disease, are more difficult to agree upon. This is due in part to the lack of consensus regarding minimal criteria for serologic diagnosis, inability to make isolates easily from infected individuals, and the lack of sensitive techniques suitable for reasonably large scale environmental sampling. The recognition of multiple *C. burnetii* genotypes of varying pathogenic potentials further complicates analysis of health within the general population [57].

However, the history of infectious disease is certainly complete with examples of serious disease causing agents, that are now regarded as relatively common, going unrecognized and unappreciated (e.g., *Legionella* & Lyme disease). Uncertainty regarding the overall incidence of Q fever and the full impact on public health, in the light of what we do know about Q fever, is to me an unacceptable argument for inaction.

Discussion: In the Q fever outbreaks associated with research facilities and hospitals, what type of environmental sampling techniques could have been used and what type of precautions should be taken in handling such samples?

Closure: In the hospital-research facility outbreaks mentioned in the text of this discussion, the source of Q fever infections was identified retrospectively by epidemiologic analysis and associated with the sheep used for medical research projects. Environmental sampling for *C. burnetii*, if the techniques for such sampling had been available, might have been used to rule out additional unidentified sources of organisms or persistence of aerosolized organisms within the indoor environment. Any environmental sampling method that depends on cultivating *C. burnetii* would definitely involve the risk of infection to those involved in cultivation of the sample. The PCR-based sample detection techniques of the near future (which will not necessarily involve cultivating live organisms) should be able to be coupled with a variety of sample collection techniques (e.g., aerosol collection either on filters or into solutions with capillary impingers). Commercial diagnostic PCR test systems for *C. burnetii* are being developed by Dr. Marvin Frazier at the Battelle Northwest Lab, Richland, WA*.

Discussion: In order to control *C. burnetii* reservoirs, should public health authorities prevent the transport of Nova Scotia cats and perhaps sheep on various forms of public transport such as large aircraft?

Closure: The Public Health Service currently makes no such recommendations for transportation of sheep or cats and it would be inappropriate for me to go on record advocating specific new policies. However, as a personal opinion, it seems reasonable to consider the possible consequences of large numbers of people sharing aerosols with potential *C. burnetii* reservoirs, especially if the animal reservoirs were parturient. On the other hand it rapidly becomes unreasonable to contemplate control of public exposure to all potential sources of *C. burnetii*.

It is likely that the Nova Scotia cats participate as amplification vehicles for a strain of Q fever perhaps found among local rodents and which may have increased potential for producing outbreaks of pneumonia in humans. It is unlikely that the Q fever reported in Nova Scotia is a disease unique only to the domestic cats of that region.

Discussion: Has *C. burnetii* been identified in animal classes other than mammals?

Closure: *C. burnetii* has long been associated with a variety of tick vectors. There are singular reports of recovery of infectious *C. burnetii* from other ectoparasites (and even flies obtained from contaminated areas) and reports of isolations from birds [48]. However, the participation of these additional arthropods and vertebrates in perpetuating cycles of disease has not been fully investigated.

The opinions expressed are those of the author presenting the paper (RLR) and do not necessarily reflect policy of the Public Health Service or the U.S. Department of Health and Human Services.

*Identification of commercial sources does not constitute product endorsement by the Public Health Service or the U.S. Department of Health and Human Services.

Harriet A. Burge

THE FUNGI

REFERENCE: Burge, H. A., "The Fungi," Biological Contaminants In Indoor Environments, ASTM STP 1071, Philip R. Morey, James C. Feeley, Sr., James A. Otten, Editors, American Society for Testing and Materials, Philadelphia, 1990.

ABSTRACT: The fungi are heterogenous organisms grouped together on the basis of structure, biochemistry, and physiology. They are classified primarily by mode of sexual reproduction, and are aerobic decay organisms. They cause human hypersensitivity, infectious, and toxic diseases. Environmental assessment can involve looking for obvious growth, or sampling either reservoirs or air. Interpretation of sampling data requires correlation with the nature and distribution of complaints.

KEYWORDS: fungi, spores, hypersensitivity, opportunistic pathogens, air sampling, reservoir sampling

I. Characterization of the agent
A. Structure/morphology

Living organisms are divided into five kingdoms [1]:

1. the Monera (bacteria and other "primitive" organisms with no organized nucleus);

2. the Protoctista (single-celled organisms with an organized nucleus);

3. the Animalia (multicellular organisms with organized nuclei, without rigid cell walls, that must obtain carbohydrates from environmental sources;

Dr. Burge is an associate research scientist at the University of Michigan Medical Center, R6621 Kresge I, Box 0529, Ann Arbor, MI 48109-0529.

4. the Plantae (multicellular organisms with organized nuclei, with cellulosic, rigid cell walls, and with chlorophyll, allowing synthesis of carbohydrates from CO_2 and water;

5. the Fungi (multicellular organisms with organized nuclei, with chitinous, rigid cell walls, that must obtain carbohydrates from environmental sources.

The basic unit of the fungal organism is a cell containing one or more nuclei, mitochondria and other membrane systems, and bound by a cell wall composed of acetyl glucosamine polymers. A few fungi are unicellular (e.g., the yeasts). However, most form long chains of cells called hyphae. A mass of hyphae is called a mycelium and constitutes the vegetative body of the fungus which carries on the activities that allow growth and reproduction. The mycelium may be colorless (appearing white in mass) or may contain melanin. This pigment allows some fungi to withstand ultraviolet radiation and grow on surfaces exposed to sunlight [1]. Fungi that cause human infectious disease can often exist in both mycelial and a unicellular yeast form. These are called "dimorphic" fungi. The conversion from mycelium to yeast is often related to temperature: at room temperature the fungi are mycelial, at body temperature they are yeast-like (Rippon 1988).

B. Life cycle/genetics

The fungi reproduce both sexually and asexually. Either (or both) kinds of spores can be borne directly on the mycelium, or mycelium may become differentiated into fruiting body tissue that contains the spores. Asexual reproduction is essentially cloning. Spores are formed that are genetically identical to the producing mycelium. Sexual spores are the result of genetic recombination. Some fungi require both spore stages to complete their life cycle (e.g., the plant rusts). Most of the fungi that contribute significantly to indoor air pollution reproduce primarily by asexual spores with adaptation to changing environments occurring through mutations and hyphal fusions. The spores themselves may be unicellular or contain up to 10 or 12 cells. Spores may be colorless and covered by mucopolysaccharides that allow them to be dispersed in water, or colored (with melanin or other pigments) and hydrophobic, insuring dispersion through the air.

C. Classification

Fungi are classified based on the mode of sexual reproduction, where known (Table 1). Using this character, they are divided into two phyla: the Zygomycota and the Dikaryomycota. The Zygomycota contains only two fungi commonly recovered from indoor environments: Mucor, and Rhizopus. The Dikaryomycota is further divided into the Ascomycotina (which contains the majority of indoor fungi including Histoplasma), and the Basidiomycotina (including mushrooms, rusts, smuts, and the opportunistic pathogen Cryptococcus neoformans).

TABLE 1 -- A summary of one classification system for the fungi based on sexual reproductive characters [1]

Phylum	Subphylum	Class
Zygomycota		Zygomycetes
Dikaryomycota	Ascomycotina	Ascomycetes
		Saccharomycetes
	Basidiomycotina	Phragmobasidiomycetes
		Teliomycetes
		Holobasidiomycetes

The many fungi for which sexual stages have yet to be discovered are classified by the pattern of asexual spore production. Those producing asexual spores directly on the mycelium are termed Hyphomycetes. Those producing fruiting bodies to contain the spores are called Coelomycetes [2]. Most indoor fungi belong to the Hyphomycetes. Within this group, the fungi are classified by the specific method of spore production. There are at least 6 different ways asexual spores are produced, and it requires considerable training, and often electron microscopy to accurately place many fungi in the appropriate group. Fortunately, simpler methods of classification involving, primarily, spore morphology have been developed that, although less accurate, are relatively straightforward to apply [3].

D. Physiology

Fungi carry on aerobic respiration (with end products CO_2 and water) in ways that are generally similar to those of plants and animals, but differ significantly in many steps. In addition to aerobic respiration, some fungi can ferment some substrates, resulting in alcohol or lactic acid rather than CO_2 production. The yeasts are well-known fermenters, being responsible for the production of beer and wine.

Fungi digest their food outside of the fungal cell. To accomplish this, they excrete enzymes into the environment. Many of these enzymes are unique to the fungi, and allow degradation of extremely resistant substances (e.g., lignin, cellulose, and polyethylene). There is probably at least one fungus that can digest any carbohydrate-containing substance [4].

In addition to enzymes, the fungi produce "secondary metabolites" that either accumulate in the environment, or are stored within the fungus. These metabolites may be involved in the pathogenesis of fungal invasive disease [5]. In many cases, the benefit of these compounds to the fungus is unknown. However, many have important effects on people. For example, penicillin is a fungal metabolite, as is cyclosporin (an immunosuppressant used to prevent transplant rejection). On the negative side, the most carcinogenic natural substance known is aflatoxin B1, also a fungal metabolite [6]. Fungi also produce many volatile compounds during active growth that can be odoriferous, and irritating [7, 8].

E. Ecology

The fungi have evolved primarily as decay organisms, and are responsible for most aerobic decay in nature. There are fungi that will utilize almost any non-living organic substrate, and a few that will invade plant and animal (including human) tissue.

Some fungi occupy very narrow ecological niches. Coccidioides immitis, for example, only occurs in arid desert soils with high salt content [9]. Cryptococcus neoformans can be abundant in dry, shaded (indoor) pigeon droppings, surviving with little competition in this alkaline, high-salt environment. Exposure to sunlight or to soil containing amoebae causes rapid elimination of this fungus [10, 11]. Histoplasma, on the other hand, also utilizes bird droppings (primarily those of starlings) but requires the presence of water and soil as well [9].

The majority of fungi are found on dead plant materials. Agricultural practices (e.g., harvesting grain after the plant dies) introduce millions of spores into the air [12]. All fungal spores found in indoor environments are ultimately derived from outdoor sources. However, when a spore with broad nutrient requirements (able to use a wide variety of substrates) encounters damp organic material indoors, it is able to germinate and grow, producing metabolites, volatile compounds, and new spores, and resulting in indoor contamination. It is important to remember that any substrate, indoors or out, that contains reduced carbon compounds and other nutrients and is damp will support the growth of some fungus.

Conditions in the indoor environment that are especially conducive to the growth of fungi are high relative humidity (which allows condensation, and absorption of water by hygroscopic materials), moisture that accumulates in appliances, and leaks and floods. In general, fungi prefer dampness rather than standing water, although some (e.g., Fusarium, Phialophora, yeasts) will grow in standing water and have been recovered from humidifier reservoirs) [13].

II. Human health effects
A. Infectious disease

Although most fungi are able to use non-living organic material (saprophytes), a few are pathogens, and will invade human tissues [14]. Fungi cause (in general) four general types of infectious disease:

1) cutaneous infections, where the fungus remains in the upper skin layers (e.g., ring worm)

2) subcutaneous mycoses, where the fungus usually remains in the subcutaneous tissues beneath the site of infection

3) the systemic mycoses, where the fungus becomes disseminated throughout the body in an otherwise normal host, and

4) the opportunistic infections, where the fungus can invade human tissues only when the hosts defenses are impaired.

The cutaneous and subcutaneous mycoses are not considered airborne diseases, and will not be discussed here. The common airborne systemic mycoses include:

1) Histoplasmosis (agent: _Histoplasma capsulatum_)
2) Coccidioidomycosis (agent: _Coccidioides immitis_)
3) Blastomycosis (agent: _Blastomyces dermatitidis_
4) Paracoccidioidomycosis (agent: _Paracoccidioides braziliensis_)

None of these fungi are routinely transmitted from indoor reservoirs. However, epidemics of histoplasmosis and blastomycosis have occurred when infected substrates have been disturbed near occupied interiors [15, 16], or when unusual situations allow growth of the organisms indoors [17, 18].

Some fungi will invade human tissues only under extraordinary circumstances (when the human immune system is compromised) and are called opportunistic pathogens [19]. The most familiar of these fungi are _Cryptococcus neoformans_, which colonizes pigeon droppings in enclosed environments [20], and _Aspergillus fumigatus_, which is a ubiquitous fungus that is one of the primary agents of decay at elevated temperatures (>40° C [21]. Actually, any fungus that is able to grow at the elevated temperature of the human body, and is able to utilize the particular compounds present can be an opportunistic pathogen. Fortunately, the human immune system is well-equipped to prevent such invasions. Only when the immune system becomes compromised does such disease occur. The primary factors that decrease immune function include disease (AIDS, some forms of cancer), and immunosuppressive medication (e.g., steroids, cyclosporin) [22].

B. Antigens/allergens

Antigens are substances that induce an immune response. The response can be cellular, or mediated by serum antibodies (immunoglobulins). Immunoglobulins can be of several different classes. Immunoglobulin G (IgG) and immunoglobulin E (IgE) are most commonly associated with responses to respiratory antigen exposure. An antigen is called an allergen when it stimulates production of IgE antibodies. Antigens are soluble, usually high molecular weight complex substances (often proteins or glycoproteins) [23].

Fungi produce a variety of these kinds of compounds, and all are potentially antigenic and allergenic [24]. In most cases, sensitization to antigens (and allergens) occurs via the airborne route. Two types of diseases that are caused by airborne fungal antigens are allergic disease (asthma and rhinitis [25]) and hypersensitivity pneumonitis [26, 27]. In addition, some fungi can grow in the thick secretions that can buildup in the lungs of some asthmatic patients. These fungi do not actually invade the human tissue, but grow in the mucus and produce antigens (and possibly toxins) that cause disease. The most common fungus causing this disease is Aspergillus fumigatus, a ubiquitous environmental fungus that is also an opportunistic infectious agent [28].

C. Toxins

Fungal toxins will be discussed elsewhere.

D. Irritants

The fungi produce volatile organic compounds during degradation of substrates that cause the typical "moldy" odor associated with fungal growth, as well as a wide variety of other odors. These substances can be irritating to the mucous membranes, and some evidence is accumulating that they may cause headaches and possibly other kinds of acute toxic symptoms [7, 8].

III. Environmental assessment

For contamination of the air to occur with fungi or fungal products, a reservoir containing living fungi is necessary to provide inoculum; an amplifier is necessary to allow growth and reproduction of the organism; and a means of dissemination is required [29]. Any assessment of the indoor environment for contamination with fungi should take these factors into account. If the presence of fungal growth in a reservoir is sufficient evidence for contamination, and a direct association with specific symptoms is not required, visual investigation is often sufficient. If documentation is required that a particular reservoir is contaminated (or amplifying the organism), samples of the reservoir material can be collected and analyzed. If one must document that airborne contamination is occurring, air sampling is required. The approach summarized below is described in more detail in [30].

A. Visual investigations

Visual investigation involves actually looking for fungal growth in reservoirs. A reservoir can be a dirty filter, damaged and wet sound lining in a ventilation system, standing water in drip pans or humidifiers, or condensation or intrusive water on surfaces [31]. The presence of slime, obvious fungal growth or obvious moldy odors can be assumed to represent contamination. Bird roosting sites adjacent to buildings should be suspected if histoplasmosis is reported [16]. Accumulated dry pigeon droppings are most likely to harbor *Cryptococcus* [20].

B. Source sampling

Source samples allow confirmation of reservoir contamination and identification of the specific fungi that are seen on environmental surfaces. Bulk samples can be collected to assess or confirm reservoir contamination. Note, however, that not all fungi will grow readily in culture. Some common reservoirs that can be assessed include those in Table 2.

TABLE 2 -- Common reservoirs/amplifiers for fungi in the indoor environment

Water reservoirs
Humidifiers [13, 32]
Cooling coil drip pans [31]
Dehumidifiers
House plant containers [33]
Fermentation vats
Water intrusion (leaks/spills) [34]
Carpeting [33]
Ceiling tiles
Wall board
Insulation
Shower curtains [35]
Flooded basements

Condensation/absorption by hygroscopic materials
Windowsills
Ventilation system ductwork [31]
House dust (human and animal skin scales) [36]
Cold (poorly insulated) wall surfaces

Organic material
Hay and other stored feed [28, 37, 38]
Stored cheese [39]
Mushroom compost
Sewage compost
Wood chips/sawdust [40, 41]
Accumulated bird droppings [42]

These samples should be collected into clean (not necessarily sterile) plastic bags (solid material) or plastic bottles (water). Analysis for fungal saprophytes involves culturing the material in ways that allow estimation of the number of microorganism per gram or square centimeter of substrate or per milliliter of water. Note that no culture method is optimal for all fungi, and many are fastidious and will not compete in culture with other organisms, or require very specialized cultural conditions. *Histoplasma*, for example, cannot be isolated directly from substrates. Cultural isolation requires inoculation of an animal host. Alternatively, a method has been described where samples are fractionated and a spore-rich fraction is examined microscopically. While not providing information on virulence, this method seems reliable for detection of probably contamination [43]. Microscopic identification can also be used for many other fungi (e.g., *Stachybotrys*).

In cases where hypersensitivity disease is of concern, bulk samples can be eluted (solid samples) or used directly (water samples) in assays designed to detect specific human antibodies directed against material in the sample [44]. In addition, bulk samples can be analyzed directly for toxins [6].

Swabs or cellophane tape samples can be taken to determine the identity of a particular contaminating fungus. Swabs are streaked onto culture medium and the predominant fungi that grow are identified. Cellophane tape can be pressed against obvious fungal growth. Microscopic examination may reveal the identity of the fungus if spores are currently being produced.

Note that fungi need not be alive and culturable to be antigenic, and toxins remain in the environment long after the fungus is dead. Methods that rely exclusively on viability may overlook serious cases of contamination [45]. Also, the presence of a toxigenic fungus (i.e., a fungus that produces metabolites with significant animal [including human] toxicity) does not imply the presence of the toxin. To be sure, one must analyze for the toxin directly.

C. Air sampling

Air sampling, in general, should only be done to confirm that specific reservoirs are producing airborne contamination. In the absence of visible reservoirs, air sampling rarely will confirm airborne fungal contamination resulting from indoor growth. In addition, it is unreasonable to expect limited air sampling to prove that contamination is not present. Many fungi either do not grow in artificial culture (i.e., the obligate parasites), or grow very slowly and are unable to compete with the very common fast growing fungi such as some *Cladosporium* and *Penicillium* species. Also, the fungi go through growth cycles where spores may not be produced, followed by periods of high spore production. Sampling during a low-spore period is likely to yield negative results.

In passively ventilated interiors, most fungal spores come in from outdoors. These, of course, will be detected by air sampling, and the levels encountered (when compared to simultaneous measurements from outdoors) can provide a measure of penetration into the dwelling, and can allow monitoring of control actions (e.g., keeping doors and windows closed) [46]. In mechanically ventilated buildings, outdoor contamination is rarely a problem.

1. Sample collection

Air samples must be collected using volumetric devices of known particle collection efficiency [47]. Fungal spores range in size from 1 um to more than 100 um. Antigens and toxins can be carried on larger particles or can be as small as a single molecule. Volatile compounds, of course, are gaseous. Most fungal spores are efficiently collected by suction impaction onto a sticky surface (grease or culture medium) [48]. Antigens, and possibly toxins carried on particles, can often be collected by filtration, using filters with small pore sizes, or consisting of materials that will absorb or adsorb the compounds of interest [49, 50]. Volatile compounds must be collected either by adsorption, or by freezing of the sample.

2. Analytical methods

The method used for sample analysis depends on the collection mode, and on the type of information desired from the sample [51]. Culture media for survey use should support the growth of as many different kinds of fungi as possible, and inhibit as few as possible. Specialized media are available for some fungi (e.g., *Cryptococcus*) [52]. Culture plate samples are usually incubated for at least 5 days at room temperature for saprophytes, 37°C for human pathogens. Near UV illumination can be used to stimulate sporulation [53]. If information on thermotolerant fungi is of primary interest, plates can be incubated at 40-45°C which will suppress most fungi, and allow Aspergillus fumigatus and some other opportunistic pathogens to grow [14]. Colonies are counted and the counts converted for multiple hole impaction if a sieve plate impactor is used [54]. Total counts are reported as colony forming units per cubic meter of air. Each sporulating colony is then identified to genus. It is useful to identify species wherever possible. Aspergillus species can usually be reported [55]. Species must be determined as well as virulence (by animal inoculation) if cause/effect relationships are to be established for infectious agents.

Impaction samples collected on optically suitable surfaces (transparent plastic or glass) can be examined microscopically, spore counts made, and some taxa identified [51, 45]. Counts are reported as spores per cubic meter of air. When combined with culture plate sampling, microscopic analysis of impaction samples allows recognition of nonviable spores. Care must be taken in comparing culture plate and microscopically analyzable samples that the relative efficiencies of the sampling devices be taken into consideration. For example, a sieve plate impactor that efficiently traps particles as small as 0.65 um in diameter is more efficient than a slit sampler that is only 50% efficient at 5 um.

Samples collected by filtration can be analyzed in a variety of ways. The filter material can be cultured directly or after elution [12]. Spore counts can be made directly on filter material (after clearing), or following elution. Although filtration samplers are very efficient at collection of particles of appropriate size (larger than the rated pore size of the filter), these methods often tend to underestimate spore levels measured by culture plate or microscopic slide impaction. Filtration samples can also be eluted appropriately and analyzed for antigens and/or toxins. It is important to understand the characteristics of the filter media being used. Some filters will irretrievably adsorb materials (see under endotoxin in this volume) so that elution is not possible. Direct analysis of the filter material may be an alternative in these cases.

Antigens are assayed using immunological methods. Most effective are monoclonal antibodies directed against the antigen of interest. Mite antigens are measured in this way [44]. The disadvantage of many monoclonal antibodies is that they may not recognize the same active sites as human antibodies. Semi-purified human antibodies are often used in immunoassays when available, but have the disadvantage of not being directed against unique antigens [56].

Toxin analysis will be covered elsewhere. Volatile organic compounds produced by fungi are analyzed in the same way as those emitted from building materials. Samples are eluted from sorbents and analyzed using gas chromatography and mass spectroscopy [57].

IV. Data interpretation

A. Establishing cause and effect

Ideally, in any investigation involving building-related symptoms, one should clearly establish that the symptoms are caused by the specific factor measured. This is almost never possible. However, a close approximation can be achieved by using the following steps:

1) Assess the symptoms and list all known potential causes.

2) If an organism or organisms have been isolated that could have caused the symptoms, establish the following:

a. that the organism was isolated from a reservoir where amplification is likely to occur and dissemination of the organism or its effluents into the breathing zone is readily hypothesized, or that the organism or its effluent was actually isolated from breathing zone air;

b. that symptomatic people were exposed, asymptomatic people were either not exposed or not at risk for symptom development.

c. that the measured exposure is the most likely cause of the symptoms (i.e., similar exposures elsewhere are less likely).

Table 3 summarizes an imaginary investigation of a fungal-related episode of building-related illness.

TABLE 3 -- Summary of an imaginary case study involving fungus-caused building-related disease

<u>Site:</u>

* 3-year-old, 4-story office building, mid-Atlantic state
* mechanical ventilation, minimal filtration, chillers
* metal duct work, no porous liners
* "as designed" (no renovation or remodeling)

Population/complaints
* ~200 office workers
* one physician-diagnosed case of hypersensitivity pneumonitis (HP)
* 5 self-reported cases of work-related asthma
* ~100 self-reported cases of nonspecific symptoms (headache, eye irritation, sinus congestion.

Walk-through observations, August
* intake filters covered with layer of greenish dirt
* sound lining in air handling unit coated with greenish dirt
* cooling coil drip pan with 1" of slimy water
* ductwork with very little obvious dirt collection
* occupied space clean, no obvious microbial growth

Sampling results
* Bulk sample of drip pan water: 10^7 gram negative bacteria/ml
* Bulk sample of sound liner from AHU: >10^6 *Penicillium chrysogenum*/gram
* Cultural air sample outdoors: >10^4 cfu *Cladosporium*; 100 cfu *Penicillium oxalicum*/m^3
* Air sample indoors, ventilation system off: no fungi recovered (<35 cfu/m^3)
* Air sample indoors, ventilation system on: 100 cfu *Cladosporium*, 75 cfu *Penicillium chrysogenum*, 50 cfu *Penicillium oxalicum*, 100 cfu miscellaneous fungi/m^3

Serological studies
* HP patient with positive precipitins to *Penicillium chrysogenum*
* one of five "asthma" patients skin-test positive to *P. chrysogenum*

Interpretation
* Air handling unit contaminated with both *Penicillium chrysogenum* and gram-negative bacteria
* Air samples confirm that occupants are potentially at risk for unusual exposure to *Penicillium chrysogenum*
* HP case exacerbated by building occupancy; no proof that primary sensitization occurred in the building.
* One asthmatic patient exacerbated by building occupancy no proof that primary sensitization occurred in the building
* Irritational symptoms may be result of exposure to volatiles produced by active *Penicillium* growth.

Recommendations
* Remove HP and *Penicillium*-sensitive asthmatic patients from the building
* Remove and replace all fiberglass lining in the AHU
* Clean and disinfect (10% bleach) the cooling coil drip pan and modify for proper drainage

B. Potential risk

Unfortunately, assessing numeric risk levels for airborne fungi is currently impossible. A few general guidelines can be proposed:

1) The indoor environment should not constitute an unusual exposure situation (either with respect to elevated levels of fungus effluents or kinds of fungus materials). Controls for comparison should include outdoor air, and comparable interiors known to have no indoor sources for fungus aerosols.

2). Fungi known to produce toxins that have serious health effects in small concentrations (e.g., *Aspergillus flavus*, *A. parasiticus*, *Stachybotrys atra*, etc.) should not be allowed to grow in non-agricultural indoor environments. Presence of these species in concentrations equal to or exceeding those in concurrently sampled outdoor air should be considered inappropriate.

3). Airborne fungus spore levels in facilities occupied by seriously immune-impaired people should be kept as low as possible, probably consistently below 100 cfu/m^3 of air as measured by an Andersen sampler on malt extract agar incubated at 30 degrees C.

However, much remains to be learned about the effects on people of saprophytic fungal growth in indoor environments. For example, symptoms of "sick building syndrome", while not generally attributed to bioaerosols, could be the result of exposure to currently unknown bioeffluents. Because of these unknowns, saprophytic growth in public interiors should be minimized. Obvious fungal growth on surfaces and in reservoirs should be considered a potential risk for disease, and a potential cause for non-specific building related complaints. On the other hand, bioaerosols should not be blamed for non-specific complaints in the absence of evidence of contamination.

REFERENCES

[1] Kendrick, B., *The fifth kingdom*, Mycologue Publications, Waterloo, Ontario, 1985.

[2] Sutton, B. C., *The Coelomycetes*, Commonwealth Mycological Inst. Kew., 1980.

[3] Barnett, H. L. and Hunter, B. B., *Illustrated genera of imperfect fungi*, 4th edition, Macmillan Publishing Co., New York, 1987.

[4] Ainsworth, G. C. and Sussman, A. S., *The fungi*, Academic Press NY, 1965.

[5] Eichner, R. D. and Mullbacher, A., "Hypothesis: Fungal toxins are involved in aspergillosis and aids," Australian Journal of Experimental Biology and Medicine, Vol. 62 (Pt. 4), 1984, pp. 479-484.

[6] Shank, R. C., Mycotoxins and N-nitroso compounds: environmental risks, CRC Press, Boca Raton FL, 1981.

[7] Kaminsky, E., Libbey, L. M., Stawicki, S., and Wasowicz, E., "Identification of the predominant volatile compounds produced by Aspergillus flavus," Applied Microbiology, Vol. 24, No. 5, 1972, pp. 721-726.

[8] Kaminsky, E., Stawicki, S., and Wasowicz, E., "Volatile flavor compounds produced by molds of Aspergillus, Penicillium, and fungi imperfecti," Applied Microbiology, Vol. 27, No. 6, 1974, pp. 1001-1004.

[9] Ajello, L., "Coccidioidomycosis and histoplasmosis. A review of their epidemiology and geographic distribution," Mycopathologia, Vol. 45, 1971, pp. 221-230.

[10] Bunting, L. A., Neilson, L. B., et al., "Cryptococcus neoformans: a gastronomic delight of a soil ameba," Sabouraudia, Vol. 17, 1979, pp. 225-232.

[11] Ishaq, C. M., Bulmer, G. S., et al., "An evaluation of various environmental factors affecting the propagation of Cryptococcus neoformans," Mycopathologia, Vol. 35, 1968, pp. 81-90.

[12] Muilenberg, M., Chapman, N., Chapman, J., and Burge, H., "Exposure of farmers, their families and townspeople to viable bioaerosols," Journal of Allergy and Clinical Immunology, Vol. 81, No. 1, 1988, pg. 274.

[13] Burge, H. A., Solomon, W. R., and Boise, J. R., "Microbial prevalence in domestic humidifiers," Applied Environmental Microbiology, Vol. 39, No. 4, 1980, pp. 840-844.

[14] Rippon, J. W., Medical mycology: the pathogenic fungi and the pathogenic actinomycetes, W. B. Saunders Co., Philadelphia, 1988.

[15] Waldman, R. J., England, A., et al., "A winter outbreak of acute histoplasmosis in northern Michigan," American Journal of Epidemiology, Vol. 117, 1983, pp. 68-75.

[16] Dodge, H. J., Ajello, L., et al., "The association of a bird roosting with infection of school children by Histoplasma capsulatum," American Journal of Public Health, Vol. 55, 1965, pp. 1203-1211.

[17] Wilcox, K. R., "The Walwort Wisconsin epidemic of histoplasmosis," Annals of Internal Medicine, Vol. 49, 1958, pp. 388-418.

[18] Denton, J. F., DiSalvo, A. F., "Isolation of Blastomyces dermatitidis from natural sites at Augusta Georgia," American Journal of Tropical Medicine and Hygiene, Vol. 13, 1964, pp. 716-722.

[19] Emmons, C. W., "Natural occurrence of opportunistic fungi," Laboratory Investigation, Vol. 11, 1962, pp. 1026-1032.

[20] Emmons, C. W., "Saprophytic sources of Cryptococcus neoformans associated with the pigeon (Columba livia)," American Journal of Hygiene, Vol. 62, 1955, pp. 227-232.

[21] Marsh, P., Millner, P., et al., "A guide to recent literature on aspergillosis as caused by Aspergillus fumigatus, a fungus frequently found in self-heating organic matter," Mycopathologia, Vol. 69, 1979, pp. 67-81.

[22] Rinaldi, M., "Invasive aspergillosis," Review of Infectious Disease, Vol. 5, 1983, pp. 1061-1077.

[23] Middleton, E., Reed, C. E., and Ellis, E. F., Allergy, principles and practice, C. V. Mosby Co., St. Louis, 1983.

[24] Burge, H. A., "Fungus allergens," Clinical Review of Allergy, Vol. 3, 1985, pp. 319-329.

[25] Salvaggio, J. and Aukrust, L., "Mold-induced asthma," Journal of Allergy and Clinical Immunology, Vol. 68, 1981, pg. 327.

[26] Fink, J. N., "Hypersensitivity pneumonitis," in Middleton, E., Reed, C. E., Ellis, E. F., Allergy, principles and practice, C. V. Mosby Co., St. Louis, 1983.

[27] Salvaggio, J. E., "Hypersensitivity pneumonitis," Journal of Allergy and Clinical Immunology, Vol. 79, No. 4, 1987, pp. 558-571.

[28] Pepys, J., "Hypersensitivity diseases of the lungs due to fungi and organic dusts," Monographs on Allergy, Vol. 4, 1969, pg. 69.

[29] Feeley, J. C., "Impact of indoor air pathogens on human health," in Gammage, R. B., Kaye, S. V., Eds., 2Indoor Air and Human Health, Lewis Publishers, Inc., Chelsea, MI, 1985.

[30] "Guidelines for the assessment of bioaerosols in the indoor environment," American Conference of Governmental Industrial Hygienists, ACGIH, Cincinnati, OH, 1989b.

[31] Morey, P. R., Hodgson, M. J., Sorenson, W. G., Kullman, G. J., Rhodes, W. W., and Visvesvara, G. S., "Environmental studies in moldy office buildings: biological agents, sources and preventive measures," Annual American Conference of Governmental Industrial Hygiene, Vol. 10, 1984, pp. 21-35.

[32] Tourville, D. R., Weiss, W. I., Westlake, P. T., and Leudemann, G.M., "Hypersensitivity pneumonitis due to contamination of a home humidifier," Journal of Allergy and Clinical Immunology, Vol. 49, 1972, pg. 245.

[33] Kozak, P. P., Jr., Gallup, J., Commins, L. H., and Gillman, S. A., "Currently available methods for home mold surveys. II. Examples of problem homes surveyed," Annals of Allergy, Vol. 45, 1980, pp. 167-176.

[34] Patterson, R., Fink, J. N., Miles, W. B., Basich, J. E., Schleuter, D. B., Tinkelman, D. G., and Roberts, M., "Hypersensitivity lung disease presumptively due to Cephalosporium in homes contaminated by sewage flooding or by humidifier water," Journal of Allergy and Clinical Immunology, Vol. 68, 1981, pp. 128-132.

[35] Green, W. F., "Precipitins against a fungus, Phoma violacea, isolated from a mouldy shower curtain in sera from patients with suspected allergic interstitial pneumonitis," Medical Journal of Australia, Vol. 1, 1972, pp. 696-698.

[36] Gravesen, S., "Identification and prevalence of culturable mesophilic microfungi in house dust from 100 Danish homes," Allergy, Vol. 33, 1978, pg. 268.

[37] Kobayashi, M., Stahmann, M. A., Rankin, J., et al., "Antigens in moldy hay as the cause of farmer's lung," Proceedings of the Society of Experimental Biological Medicine, Vol. 113, 1963, pg. 472.

[38] Gregory, P. H. and Lacey, M. E., "Mycological examination of the dust from mouldy hay associated with farmer's lung disease," Journal of General Microbiology, Vol. 30, 1963, pg. 75.

[39] Campbell, J. A., Kryda, M. J., Treuhaft, M. W., Marx, 3J. J., Jr., and Roberts, R. C., "Cheese worker's hypersensitivity pneumonitis," American Review of Respiratory Diseases, Vol. 127, 1983, pg. 495.

[40] Van Assendelft, A. H., Raitio, M., and Turkia, V., "Fuel chip-induced hypersensitivity pneumonitis caused by Penicillium species," Chest, Vol. 87, 1985, pg. 394.

[41] Emanuel, D. A., Wenzel, F. J., and Lawton, B. R., "Pneumonitis due to Cryptostroma corticale (maple bark disease)," New England Journal of Medicine, Vol. 274, 1966, pg. 1413.

[42] Ruiz, A., Neilson, J. B., and Bulmer, G. S., "A one year study on the viability of Cryptococcus neoformans in nature," Mycopathologia, Vol. 77, 1982, pp. 117-122.

[43] Gaur, P. K. and Lichtwardt, R. W., "Preliminary visual screening of soil samples for the presumptive presence of Histoplasma capsulatum," Mycologia, Vol. 72, No. 2, 1980, pp. 259-269.

[44] Platts-Mills, T. A. E., Chapman, M. D., Heymann, P. W., and Luczynska, C. M., "Measurements of airborne allergen using immunoassays," in Solomon, W. R., Ed., Immunology and Allergy Clinics of North America. Vol 9, WB Saunders, Philadelphia, 1989, pp 269-283.

[45] Burge, H. P., Boise, J. R., Rutherford, J. A., Solomon, W. R., "Comparative recoveries of airborne fungus spores by viable and nonviable modes of volumetric collection," Mycopathology, Vol. 61, No. 1, 1977, pg. 27.

[46] Chatigny, M. A., Dimmick, R. L., "Transport of aerosols in the intramural environment," in Aerobiology: the Ecological Systems Approach, Edmonds, R. L., Ed., Dowden, Hutchinson & Ross, Stroudsburg, PA, 1969.

[47] "Air sampling instruments for evaluation of atmospheric contaminants," American Conference of Governmental Industrial Hygienists, 7th edition, ACGIH, Cincinnati, OH, 1989a.

[48] Burge, H. A. and Solomon, W. R., "Sampling and analysis of biological aerosols," Atmosphere and Environment, Vol. 21, No. 2, 1987, pp. 451-456.

[49] Hoyer, M., Arscott, P., Burge, H., Klapper, D., Solomon, W., and Baker, J., Jr., "Detection and quantitation of airborne ragweed allergens using a new immunoblotting technique," Journal of Allergy and Clinical Immunology, Vol. 85, No. 1, Pt. 2, 1990, pg. 227.

[50] Jensen, J., Poulsen, L. K., Mygind, K., Weeke, E. R., and Weeke, B., "Immunochemical estimations of allergenic activities from outdoor aeroallergens, collected by a high volume air sampler," Allergy, Vol. 44, No. 1, 1989, pp. 52-59.

[51] Muilenberg, M. L., "Aeroallergen assessment by microscopy and culture," in Solomon, W., Ed., Immunology and Allergy Clinics of North America, Vol. 9, No. 2, 1989, pp. 245-268.

[52] Racicott, T. A., Bulmer, G. S., "Comparison of media for the isolation of Cryptococcus neoformans," Journal of Applied Environmental Microbiology, Vol. 50, 1985, pp. 548-549.

[53] Burge, H. P., Solomon, W. R., Boise, J. R., "Comparative merits of eight popular media in aerometric studies of fungi," Journal of Allergy and Clinical Immunology, Vol. 60, No. 3, 1977, pp. 199-203.

[54] Andersen, A. A., "New sampler for the collection, sizing and enumeration of viable airborne particles," Journal of Bacteriology, Vol. 76, 1958, pp. 471-484.

[55] Raper, K. B. and Fennell, D. I., "The genus Aspergillus," 1965, available as reprint from R. E. Krieger Publishing Co., Malabar, FL.

[56] Swanson, M. C., Agarwal, M. K., and Reed, C. E., "An immunochemical approach to indoor aeroallergen quantitation with a new volumetric air sampler: studies with mite, roach, cat, mouse, and guinea pig antigens," Journal of Allergy and Clinical Immunology, Vol. 76, No. 5, 1985, pp. 724-729.

[57] Lewis, R. G. and Wallace, L. P., "Toxic organic vapors in indoor air," ASTM Standardization News, December 1988, pp. 40-44.

[58] Morse, D. L., Gordon, M. A., Matte, T., and Eadie, G., "An outbreak of histoplasmosis in a prison," American Journal of Epidemiology, Vol. 122, 1985, pp. 253-261.

DISCUSSION

How do you interpret your results (e.g., from the Andersen sampler) when you have 2 very different plate count results? What guidelines do you follow to do your interpretation?

CLOSURE

If one plate is blank, the other with many (>10 colonies), we use only the positive plate. If both plates have reasonable numbers of colonies, we use either an average or a range depending on the type of investigation.

DISCUSSION

In a recent American Industrial Hygiene Conference in St. Louis, a Canadian investigator stated that 50 or more _Stachybotrys_ per cubic meter or 300 or more total fungi per meter cube is a concern level. Please comment.

CLOSURE

Fifty or more _Stachybotrys_/m^3 would prompt me to look for a source of _Stachybotrys_ in the environment. In normally ventilated homes, 300 cfu total fungi/m^3 cannot be considered unusual. I do not feel an overall concern level can be set for total fungi. The environment, the population at risk, and the sampling modality must be a part of any guideline.

DISCUSSION

Are there problems in comparing outside sample fungi results vs indoor findings? (i.e.: varying normal flora etc)? How do you compare the results?

CLOSURE

Significant problems arise. 1) Sampler efficiency may change depending on wind speed. 2) The air spora may be differently (spatially) distributed indoors vs outdoors 3) Usually, generic identifications are used for such comparisons. Unless only one species occurs, such comparisons are not valid. We compare total spore counts indoors and out (assuming that total fungus exposure is meaningful with respect to disease). When looking at particular types of fungi (e.g., _Penicillium_) we identify each isolate to species.

DISCUSSION

Have you had any experience with the Biotest Centrifugal Air sampler?

Can the RCS centrifugal sampler be used successfully?

CLOSURE

Yes. It is a portable sampler that is useful especially for monitoring bacterial levels in places where larger samplers are inconvenient. However, 1) it cannot be field calibrated and the manufacturers' stated "effective volume" is apparently inaccurate; 2) the agar strips are not available with malt extract agar; 3) fungi rapidly fill the strips making enumeration difficult. Note that a new model that can be calibrated is currently being tested.

DISCUSSION

What about the "nut" in the building who is [not] really getting ill from molds; how do you determine the "nut" from the sick individual who is really ill because of the exposure?

CLOSURE

Sometimes you can't make the distinction. However, most illness that is clearly related to fungus exposure produces an immune response (either IgE or IgG) that can be detected serologically. Irritant or toxic symptoms should be clearly related to building occupancy. Often the unrelated cases will have a long history of similar problems in unrelated environments.

DISCUSSION

Do you recommend checking the calibration on cascade impactors for calculated versus actual size separation? Methods? Do you ever make mass measurements from impactor stages? If so, for what application?

CLOSURE

We do very little with size separation. Most fungus spores apparently travel as units and the known spore size tells us the aerosol size. Also, allergens are effective over a very large size range. We occasionally use the 6-stage Andersen, but trust the extensive characterization work done on this sampler by the Defense Department. We always calibrate the flow rate through the device. We have not made mass measurements from impactor stages.

DISCUSSION

Could you elaborate on the significance of non-viable proteinaceous fragments in onset of allergic response, especially in atopic and/or sensitized individuals? How do you identify these?

CLOSURE

Antigen fragments are an effective means of inducing symptoms in an appropriately sensitized subject. The only way to evaluate the concentrations of such particles is by immunoassay (e.g., RAST, ELISA, Immunoblotting).

DISCUSSION

Please elaborate on thermophilic actinomycetes sampling/analysis.

CLOSURE

Thermophilic actinomycetes can be assessed by culture plate sampling using a sampler (such as the Andersen) that efficiently collects 1um spores. TSA or nutrient agars are acceptable; they must be incubated at ~56°C. If appropriate antibodies are available, these antigenic units can be assessed by filtration sampling followed by immunoassay.

DISCUSSION

If rose bengal agar needs to be incubated in the dark, isn't this medium ineffective for identifying fungi that require light for spore production?

CLOSURE

Yes.

DISCUSSION

When you have enumerated fungal spores on sticky slides, do the numbers agree with colony forming units? Are there "non-culturable fungi"?

CLOSURE

There is usually a positive and significant correlation between spore trap and suction culture plate samples. However, culture plate sampling always estimates actual levels for culturable taxa because all airborne spores are never alive, and many are stressed and do not produce colonies in artificial culture. In addition, many spores do not germinate or grow in artificial culture and are not recovered by culture plate sampling. Most of the mushroom spores and all of the obligate plant pathogens fall into this category. Mushroom spores, by the way, are highly allergenic.

DISCUSSION

How much will speciation of all fungi (as you suggested should be done) increase the cost of analysis and time required for analysis of volumetric bioaerosol samples?

CLOSURE

Speciating all fungi would be ideal but impossible. Some fungi are relatively easy to speciate (e.g., Aspergillus). Others, including the common genus Penicillium are very difficult and expensive. I estimate that it costs us at least $150 per isolate to identify most Penicillium species, and it takes a minimum of 7 days to do the necessary cultures. With limited staff, it would probably double the time required for analysis.

DISCUSSION

What biocides are appropriate for use in HVAC systems while buildings are occupied?

CLOSURE

None can be used in an operating system so that occupants are at risk for exposure to the biocide.

DISCUSSION

What modifications (if any) can be made to the RCS/SAS samplers to improve the ~40% collection efficiency relative to the Andersen for viable organisms?

CLOSURE

The relative efficiency for the RCS is not known. For the SAS, decreasing the diameter of the sieve plate holes appears to increase efficiency, although it lowers sampling rate.

DISCUSSION

Is the reduced collection efficiency of the Burkard (indoor) for 5um particles due to the "relatively" low sample flow rate or some characteristic of the sampling head?

CLOSURE

It is probably due to the width of the slit.

DISCUSSION

Is there any concern for "sample shock" or "organism shock" when utilizing the high flow rates of the RCS/SAS for viable organisms?

CLOSURE

The actual speed with which the organisms strike the agar surface is not greater than that for the Andersen, especially if you use the sixth stage. The very small hole size causes the velocity of the air stream to be quite high. Probably some fragile organisms are killed or damaged by the impact.

DISCUSSION

Under what conditions is "aggressive" sampling, i.e., pounding on ductwork or fan unit filters, recommended, and is this practice not supporting artificial recoveries if the aggressive activity does [not] coincide with some similar workplace event, i.e., maintenance?

CLOSURE

Aggressive sampling constitute a "worst-case" situation, and can be very valuable in determining whether or not a particular reservoir/amplifier can (under any circumstances) contribute to the aerosol. Usually, aggressive sampling is done in ways that attempt to duplicate activities that might actually occur in the space.

DISCUSSION

Please comment on criteria utilized for establishing typical indoor/outdoor ratios and their applicability to data interpretation.

CLOSURE

Indoor/outdoor ratios are usually calculated on the basis of total particulate counts. It is usually assumed that if indoor levels are lower than those outdoors, then the indoor environment is not contaminated with respect to outdoors. Obviously, this is not a valid assumption. Unless the particles are of identical types indoors and out, and unless patterns and rates of penetration are known, one cannot assume an outdoor source for indoor aerosols.

DISCUSSION

We sometimes use allergen skin tests on building-responding patients. When the test material is marked for example "Cladosporium," is positive response useful in light of the difficulty in identifying species?

CLOSURE

A strong positive response to a fungal allergen indicates that the individual is probably atopic. How the atopy relates to building occupancy is much more difficult to determine. Fungal allergen extracts available for skin testing are extremely variable, poorly characterized, and often mis-identified. The only way to clearly make a connection with building exposure, one would have to use for skin testing material made from a strain isolated in the building, and demonstrate that the reaction is unique to that strain. Especially with Cladosporium (which is always abundant outdoors), it is extremely unlikely that sensitization would occur only in connection with an indoor environment.

DISCUSSION

Is the agar to sieve plate distance critical for the Andersen sampler? If so, what should the distance be?

CLOSURE

Yes. The distance should be 2.5mm. For the original cast aluminum sampler using Andersen's original glass petri dishes, 27 ml of agar was necessary to achieve this distance. If plastic plates are used in this sampler, 47 ml of agar is necessary. Newer models currently available (the "Flow-sensor" samplers and the lathe-turned Andersens) have been modified for use of plastic petri dishes with 20 ml agar.

DISCUSSION

Using the N6 procedure, the average velocity per jet is 76.4 ft/sec (pretty high!). Has there been any research on whether this impaction velocity kills any important organisms?

CLOSURE

All the research has been with bacteria, and certainly some fragile organisms are killed. No studies have been done with fungi. Actually, the 5th stage would be just as efficient at collecting fungus spores (none of which are smaller than 1um) and the velocity would be only 42 ft/sec.

DISCUSSION

What types and concentrations of pathogenic fungi (Cryptococcus, Histoplasma, etc.) occur in abandoned buildings where large amounts of pigeon fecal material are being removed with and without prior formalin disinfection? Has anyone done this type of work? Assume volumetric sampling. What sampler was used?

CLOSURE

The fungus pathogen that is most likely to occur in indoor pigeon fecal material is Cryptococcus. Airborne levels have been measured (in a limited way) using the Andersen sampler by Ruiz et al. [42]. However, they did not use formalin disinfection. Reports of the use of formalin disinfection have been related to Histoplasma in outdoor sites [58]. Histoplasma cannot be evaluated using air sampling, so the data you request are unavailable.

DISCUSSION

If use of animals in the laboratory is prohibited, are there any methodologies for direct culture available (even though less sensitive) (e.g., described in the literature or possibly used by e.g., Fort Detrick?

CLOSURE

A visual screening method is available for Histoplasma (see sampling discussion above). Blastomyces is rarely isolated in connection with epidemics, even using animal inoculation procedures. Cryptococcus and other opportunistic pathogens (e.g., Aspergillus) can be isolated culturally using selective culture media or incubation conditions.

DISCUSSION

When do you feel would be a situation that air sampling investigation/evaluation should be conducted for the presence of [pathogenic] yeasts in the indoor environment?

CLOSURE

I would not recommend air sampling for these organisms. If disease exists, and a reservoir containing the organism exists, then air sampling is probably not necessary. If disease is not present, or if an apparent reservoir is not present, air sampling is unlikely to be of use. Air sampling cannot be effectively done for Histoplasma or Blastomyces.

DISCUSSION

What is the significance, if any, of the presence of Cryptococcus neoformans in an office? Assume no disease.

CLOSURE

Cryptococcus neoformans is an opportunistic pathogen and only causes infection if inhaled in very high numbers, or in immunocompromised people. The presence of the organism in an office environment means that a reservoir exists somewhere in the environment, and that the potential for disease exists if an immunocompromised person occupies the environment.

DISCUSSION

Can you suggest a good reference for decontamination of an area covered with pigeon and bat droppings?

CLOSURE

Reported decontamination methods range from treatment with formalin [58] to drying and waiting [42]. Other possibilities include the use of other biocides (such as dilute hypochlorite or dilute hydrogen peroxide) that do not carry the potential risks of formaldehyde.

DISCUSSION

Are there documented cases of disease from workers exposed to pigeon and bat droppings?

CLOSURE

Many. Reviews of the literature are included in specific chapters on histoplasmosis, blastomycosis, and cryptococcosis in Rippon's textbook of medical mycology [14].

DISCUSSION

With the tremendous capsule surrounding the yeast cell do you feel that in using an impaction device (SAS, Andersen) with quiescent and aggressive sampling techniques you could collect viable organisms?

CLOSURE

Cryptococcus usually does not have the large capsule in saprophytic environments. Viable organisms of _Cryptococcus_ can be collected using impaction sampling.

DISCUSSION

What would be the best mode of transport from the field to the laboratory (e.g., room temperature, ice pack, etc) Presume _Histoplasma_, _Cryptococcus_).

CLOSURE

Room temperature would be acceptable for _Cryptococcus_ samples. The organisms have been shown to survive for up to 9 months in a dry condition without input of new substrate. _Histoplasma_ has been less well studied. For long periods of transit (more than 24 hours), I would suggest maintaining the sample at approximately the temperature at which it was collected.

DISCUSSION

Due to the potential liability problems in recommending a formalin decontamination, what are other methods of decontamination that do not involve formalin or other materials of similar health risks?

CLOSURE

Drying and prevention of new substrate additions has been suggested for Cryptococcus, although only 85% of spores were dead after 9 months. Other biocides could probably be used (such as dilute hypochlorite or dilute hydrogen peroxide).

DISCUSSION

Is there a hazard of concern when pigeons roost on air intakes of building HVAC systems? How should it be controlled?

CLOSURE

Organic material building up in air intakes is certainly a concern, though not established as a health risk. Downstream filtration, if properly installed and maintained, should capture the majority of biological contaminants that would occur in such material. Methods for excluding pigeons is beyond the purview of this conference. Pigeon droppings, once they have accumulated, can be decontaminated by soaking with dilute formalin (or probably, dilute hypochlorite or peroxide).

Richard L. Tyndall and Kevin S. Ironside

FREE-LIVING AMOEBAE: HEALTH CONCERNS IN THE INDOOR ENVIRONMENT

Reference: Tyndall, R. L., and Ironside, K. S., "Free-Living Amoebae: Health Concerns in the Indoor Environment", Biological Contaminants In Indoor Environments, ASTM STP 1071, Philip R. Morey, James C. Feeley, Sr., James A. Otten, Editors, American Society for Testing and Materials, Philadelphia, 1990.

ABSTRACT: Free-living amoebae are the most likely protozoa implicated in health concerns of the indoor environment. These amoebae can be the source of allergic reactions, eye infections or, on rare occasions, encephalitis. While too large to be effectively aerosolized, free-living amoebae can support the multiplication of pathogens such as *Legionella* which are easily aerosolized and infectious via the pulmonary route. Traditional detection methods for free-living amoebae are laborious and time consuming. Newer techniques for rapidly detecting and quantitating free-living amoebae such as monoclonal antibodies, flow cytometry, gene probes, and laser optics have or could be employed.

KEY WORDS: free-living amoebae, *Naegleria*, *Acanthamoebae*, *Hartmannella*, *Legionella*, monoclonal antibodies, flow cytometry, gene probes, laser optics

INTRODUCTION

Unlike bacteria and viruses, protozoa have a true nucleus and are thus eukaryotes. Protozoa may range in size from 2μm to several centimeters, may be uninucleate or multinucleate and often have locomotor organelles such as flagella, cilia and pseudopodia. They generally reproduce by binary fission preceeded by nuclear division.

Richard Tyndall is a staff scientist with the Health and Safety Research Division, Oak Ridge National Laboratory, P. O. Box 2008, Oak Ridge, TN 37831; Kevin Ironside is a staff member of the Zoology Department, University of Tennessee, Knoxville, TN 37996-0810.

This research was sponsored by the Office of Health and Environmental Research, U.S. Department of Energy, under Contract DE-AC05-84OR21400 with Martin Marietta Energy Systems, Inc.

Dissolved foods are available to the organisms by pinocytosis or by simple diffusion whereas particulate food is ingested or phagocytosed through cytostomes.

There is a complex spectrum of sizes and shapes extant in the community of nonpathogenic protozoa. Various nonpathogenic protozoan species commonly inhabit mammals, including man. Conversely, some protozoa are pathogenic for man.

Of the various protozoa pathogenic for humans, the unicellular protozoa may pose the most problems in developed countries. *Entamoebae histolytica* and *Giardia lamblia* are obligate gastrointestinal parasites. Generally, however, these are not normally indoor contaminants.

Protozoans posing a greater potential problem in the indoor environment may be free-living amoebae such as *Naegleria, Acanthamoeba* and *Hartmannella*. Some species of these genera can be pathogens per se, allergenic or amplifiers of bacterial pathogens such as *Legionella*.

Description and Classification

Free-living amoebae are ubiquitous in soil and water. They are relatively small (~8 x 40 μm) and are capable of utilizing gram-negative bacteria or dissolved organic material as food sources [1,2]. Classification of the free-living amoebae based solely on morphology of the trophozoite, cysts, and flagellate forms is not totally practical because of the amorphous nature of the microbes. Nevertheless, attempts to classify these amoebae by their microscopic appearance have been undertaken. Page [3] divided the order Amoebida into a variety of families, with the genus *Naegleria* assigned to *Vahlkampfiidae*, *Hartmannella* to *Hartmannellidae*, and *Acanthamoebae* to *Acanthamoebidae*. At present, the classification suggested by Page is the most widely accepted.

In unfixed preparations the *Naegleria* trophozoites appear slug-like, the anterior end being broader than the posterior end. They move by means of eruptive pseudopodia. In addition to the trophozoite form of *Naegleria*, the microbe can also exist in a flagellate and cyst form. Transformation from the trophozoite to the flagellate form most often occurs when the supporting medium is diluted with water. The rapid motility of this form is by means of two to four anterior flagella. In the absence of nutrients, *Naegleria* can also encyst. The spherical cysts are 7 to 15 μm in diameter. Excystation occurs when suitable food sources are again available. *Hartmannella*, like *Naegleria*, are also limacine amoebae, but unlike *Naegleria*, do not form flagella. Their spherical cysts are somewhat smaller than *Naegleria*. *Acanthamoebae* are characterized by spike-like cytoplasmic projections. The amoebae generally contain a single nucleus with a centrally located nucleolus. Multinucleate forms have been seen occasionally. Unlike *Naegleria*, they do not form flagella but can encyst when deprived of nutrients. The cysts are composed of both endocysts and ectocysts.

In order to avoid the ambiguities of morphologic and cytologic classification, serologic techniques were used to provide a more reliable parameter for speciating the amoebae. By preparing hyperimmune antisera, and more recently monoclonal antibodies, against various amoebae, detailed fluorescent antibody and immunoelectrophoretic analyses of the isolates are possible. By these methods, together with isoenzyme analysis, various *Naegleria* and *Acanthamoebae* can generally be speciated [4-8].

Pathogenicity

Since the first reports concerning the isolation of potentially pathogenic free-living amoebae in 1957, free-living amoebae of the genera *Naegleria*, i.e. *N. fowleri*, and *Acanthamoeba* spp. have been recognized as the etiological agents of primary amoebic meningoencephalitis (PAME), chronic meningoencephalitis, and serious eye and wound infections. Over 50 cases of PAME have occurred in the United States [9].

Human infections with *N. fowleri* generally occur when water infested with the microorganism is exposed to the nasal mucosa of a susceptible individual. Most cases occur in persons engaged in active water sports such as swimming and diving. One report indicated that *Naegleria* encephalitis occurred after airborne exposure to the amoebic cysts (2). The amoebae traverse the nasal mucosa and migrate along the olfactory nerves through the cribriform plate to the cerebrum. The ensuing infection results in high fever, severe headache, erratic behavior and death. While a few individuals have survived infection with *N. fowleri*, most succumb in spite of extensive antibiotic therapy.

In addition to overt infection which is a rare event, exposure to *Naegleria* antigens can also elicit allergic reactions. Several studies showed that "humidifier fever" in office and industrial workers was most likely attributable to aerosolized *Naegleria* antigens [10,11].

Acanthamoebae also can cause infections. Such infections may occur after trauma and concomitant introduction of the amoebae into the traumatized site, which, in some cases, may also result in chronic, fatal encephalitis [12]. One of the documented sites of human *Acanthamoeba* infections is the eye [13,14]. *Acanthamoeba* introduced into the eye at the time of traumatization have caused eye infections leading to the loss of the infected eye. The severity of *Acanthamoeba* eye infections may be compounded, in part, by the relative ineffectiveness of most antibiotic therapy in treating such diseases.

Hartmannella, while not generally pathogenic per se can impact on human health by their propensity for supporting the growth of bacteria such as *Legionella*. This ability was first demonstrated in laboratory experiments with *Naegleria* and *Acanthamoebae* [15] and subsequently with *Hartmannella* [16]. This has also been recently demonstrated in hot water heaters where *Legionella* presence has been undeniably linked to the presence of *Hartmannella* [16].

Ecology

The ability of pathogenic *Acanthamoebae* and *Naegleria* to grow at higher temperatures than many nonpathogenic species has been demonstrated in laboratory studies [17]. This has proven to be a useful observation in its application as a semiselective technique for the isolation of pathogenic free-living amoebae from environmental samples. Water concentrates or detritus can be placed on nonnutrient agar plates seeded with a lawn of *E. coli* and incubated at 43 to 45°C. In general, pathogenic amoebae will grow more readily under these conditions relative to their nonpathogenic counterparts.

While not yet shown environmentally for *Acanthamoeba* or *Hartmannella*, with the possible exception of hot water heaters, there is a clear association between the presence of *Naegleria* spp. and warm water [18]. A detailed study of the distribution of thermophilic and pathogenic *Naegleria* spp. in thermal discharges of electric power plants in northern and southern regions of the United States relative to ambient source waters has been documented. Of the 385 samples from 17 cooling systems, 195 were positive for thermophilic amoebae (51%). Of the 195 thermophilic amoeba isolates from the power plant cooling systems, 85 (44%) were identified as *Naegleria* spp. by the flagellation test and from morphological criteria. Of the 85 identified, 16.5% were *N. fowleri*, whereas the remaining isolates were probably *Naegleria lovaniensis* [18].

In contrast to results from the cooling waters, in unheated control waters only 5% of the amoeba isolates were *Naegleria* spp., and none of these were pathogenic. Data from 17 test sites supported the association between thermophilic, pathogenic amoebae and artificially heated water. The amplification of thermophilic amoebae, presence of thermophilic *Naegleria* spp., and isolation of *N. fowleri* in heated versus unheated systems were all significant at the $P<0.05$ level by chi-square analysis.

An artificially heated lake in Virginia was the first site studied before and during thermal enrichment. The presence of *N. fowleri* was related to the initiation of thermal additions to the lake. The pathogen was not isolated before thermal additions; however, *N. fowleri* was isolated after thermal additions began. Increased numbers of amoeboflagellates growing at 44°C were also isolated [19].

Similar data were reported for two northern impoundments sampled in 1983. One site had received thermal additions from an electric power plant since 1970, and water temperatures at the sampling locations ranged from 34.8 to 41.5°C. The percentages of water and detrital samples that tested positive for the presence of thermophilic amoebae, thermophilic *Naegleria* spp., and *N. fowleri* were 100, 76, and 41, respectively [18].

The second study site was an unheated lake approximately 185 mi (ca. 298 km) from the heated site. The percentages of samples positive for thermophilic amoebae and thermophilic *Naegleria* spp. were 57 and 27, respectively. *N. fowleri* was not isolated from this site. Statistical comparison between the heated and unheated sites showed significant differences in the presence of thermophilic amoebae, thermophilic *Naegleria* spp., and *N. fowleri* at the $P<0.01$ level.

In addition to industrial sites, swimming pools, natural hot springs and solar heated ponds have also provided habitats for the survival and growth of *N. fowleri* [20-23]. Pathogenic *Naegleria* spp. were isolated from various naturally heated ponds in South Carolina. In studies by Kyle and Noblet [24], *N. fowleri* was shown to be associated with particulate detritus layers in such ponds. An analysis of various thermal springs in New Zealand for *N. fowleri* relative to a variety of habitat parameters was undertaken in 1976 and 1977 [20]. Sample sites were generally those associated with previous cases of human meningoencephalitis. Of the total samples tested in 1976, 23% were positive.

A survey of heated and ambient habitats at the Savannah River plant (SRP) in Aiken, S.C., for thermophilic amoebae in general and thermophilic *Naegleria* spp. in particular was conducted in 1976 [25]. Thermophilic *Naegleria* spp. were detected 43% of the time in thermally altered habitats but only 2% of the time in ambient-temperature habitats. The data suggested that thermally altered aquatic systems in the southeastern United States may provide habitats conducive for proliferation of the thermophilic and pathogenic *Naegleria*.

Isolation and Detection

All of the above studies were carried out using methodologies developed decades ago. These assays for pathogenic *Naegleria* are both labor intensive and require weeks to detect the presence of pathogenic amoeba.

The traditional method for detecting free-living amoebae involves plating a mixture of environmental microbes on nonnutrient agar plates spread with a lawn of *E. coli* (NNAE) and incubating the plates at 25°, 37° or 45°C. The plates are then observed microscopically for outgrowth of amoebae. Analyses for amoebic populations can be quantitative, qualitative or both. For most probable number determinations, five replicates each of 0.01-, 0.1-, 1.0-, 10-, and 100-ml water samples are plated. The 100-ml water samples are filtered through 1.2-μm-pore-size cellulose filters which are then quartered or halved and inverted onto the NNAE plates. The 10-ml samples are centrifuged at 600 x g for 15 min, and the resultant pellets are plated on NNAE plates. The remaining water samples are placed directly onto the surfaces of the NNAE plates.

For detecting and isolating pathogenic *Naegleria*, the NNAE plates are incubated at 45°C for 7 days or until growth of thermophilic amoebae is observed. These initial outgrowths determine the concentration of thermophilic amoebae. Morphological characteristics of the trophozoite and cyst are used to tentatively identify amoebae belonging to the genus *Naegleria*, *Acanthamoeba* or *Hartmannella*.

All thermophilic *Naegleria* organisms that grow at 45°C with morphology patterns indicative of *Naegleria* spp. are then tested for the ability to flagellate. The flagellation tests are performed at 35°C. After being harvested, the trophozoites are suspended in sterile distilled water and examined under an inverted microscope for the presence of flagellates after 1, 2, and 3 h. Results of these analyses determine the concentration of thermophilic *Naegleria* spp.

To determine the concentration of pathogenic *N. fowleri*, the thermophilic *Naegleria* isolates are washed and suspended in sterile distilled water. The suspension (4,000 to 10,000 amoebae) is intranasally instilled in weanling BALB/c mice. Test mice are observed for 2 weeks for signs of encephalitis. Moribund mice are sacrificed, and their brain tissue is plated on NNAE plates. Outgrowth of *Naegleria* spp. from the tissue confirms the presence of pathogenic *N. fowleri*. Negative controls consist of weanling mice inoculated with nonpathogenic amoebae, necropsied, and processed as described above.

Presence and quantitation of thermophilic or thermotolerant *Acanthamoeba* or *Hartmannella* in the samples can be made by allowing the trophozoites on the test NNAE plates to encyst. The size and double walled morphology of *Acanthamoeba* cysts allows for their detection and quantitation, whereas the morphology, size and single layered cyst wall will distinguish the *Hartmannella*. For detection of less thermotolerant species, replicate plates of the various concentrations of samples should be plated at 35°C. Pathogenicity tests for encephalitic *Acanthamoeba* populations are carried out as described for pathogenic *Naegleria*.

While extensive analysis of thermally altered waters have amply demonstrated the association between thermal additions and pathogenic *Naegleria*, few studies of airborne amoebae or amoebic antigens have been made. One report does indicate possible human infection with airborne *Naegleria* cysts [26]. However, air samples taken proximal to water with $\geq$ 20,000 *Naegleria* per liter did not detect airborne *Naegleria* [27]. Other studies indicate allergic reactions in workers exposed to aerosolized *Naegleria* antigens.

Air Sampling

Because of the large size of trophozoites and cysts and the water droplets required for their aerosolization, it would be expected that the distance traversed and duration of flight from their point of aerosolization should be minimal. Consequently relatively large samples taken close to the point of aerosolization are recommended. Three types of air samplers can be used for sampling. These include impingers, Andersen samplers, and high volume samplers such as the Litton and cyclone samplers [28].

The Sci-Med Model M-3A Litton-type high-volume air sampler is designed to collect particulate matter continuously from a large air sample (maximum of 1 x 10^3 L/min) and deposit it into a small amount of liquid (300 mL). The sampler uses electrostatic precipitation to separate microorganisms from air. This effects a maximum concentration factor on the order of 1.8×10^5 during a 3-h sampling period.

The Greenberg-Smith impinger air sampler is a simple vacuum-driven device capable of processing air at approximately 25 L/min. The "inhaled" air is passed through sterile filtered water so that the amoebae remain in the water phase. A second reservoir is connected in series with the first; most amoebae inadvertently carried through the first reservoir will be collected in the second. After sampling for 30- to 60-min, the two 150-mL water reservoirs are combined and analyzed for amoebae. The flow rate of the air sampler is set using a calibrated rotameter.

Microorganisms collected by sampling with either the high volume or impinger samplers are concentrated by centrifugation and tested for the presence of amoebae as previously described or by newer technologies currently under development.

When the Anderson sampler is used, air is passed through the sixth stage plate. The metal plate has 400 holes through which the airborne microbes pass before being deposited on the surface of a nonnutrient agar plate spread with a lawn of *E. coli*. Air is sampled at a rate of 28 L/min for ten to thirty minutes. Since any collected amoebae may migrate from the point of impact, amoebic colonies have to be observed daily to ensure that satellite colonies did not originate from original colonies arising from the point of impact. For collecting air samples, the samplers are placed such that they would not be inundated by direct exposure to falling water but would be close enough to sample mist.

As previously discussed, most tests for pathogenic free-living amoebae still use the time honored, but time consuming and costly plating and animal assays for the presence of pathogenic free-living amoebae. Only in the past few years have newer technologies been used in environmental testing for pathogenic protozoa.

To comprehend how recently developed technologies such as monoclonal antibodies, flow cytometry fiber optics and gene probes have or may facilitate the rapid detection and quantitation of protozoan or their antigens, we will consider past outbreaks and consider how these could be handled in future studies.

Case Studies

In 1976 an outbreak of fever in workers in an industrial setting prompted an investigation into the possible involvement of *Naegleria*. Air sampling was conducted using the Anderson impact sampler fitted with nonnutrient agar plates spread with a lawn of *E. coli*. After many days of incubation, amoebic outgrowth was observed on some plates. Morphologic analyses indicated that the amoebae were *Naegleria* [29]. The amoebae were then isolated and extracts prepared. Serologic tests were used to show that the isolates were indeed *Naegleria* and may have caused worker illness. The source of the amoebae was then traced to the humidification system where successful remedial action was undertaken.

An epidemic of 16 deaths from meningoencephalitis between 1962 and 1965 in Czechoslovakia was reported by Cerva. After three deaths in 1962 and four in 1963 were traced to an indoor swimming pool, the pool was closed and cleaned. Additional cases appeared after the pool was reopened in 1964, and within days five people became ill and died. After additional analyses, the source of the causative agent, *N. fowleri*, was found sequestered behind repairs previously made on the wall of the swimming pool [30].

Future analysis of indoor environments for suspected protozoa or their antigens, as in "humidifier fever" and swimming pool scenarios just described, will hopefully employ more rapid and definitive techniques so that prompt remedial action can be taken.

One of the main tools for detection of intact, viable protozoa or antigenic material released from the protozoa is, and will be, monoclonal antibodies [31]. Monoclonal antibodies prepared against free-living amoebae of interest will be prepared and stored until needed. While the initial monoclonal antibody production per se is a labor intensive and costly procedure, once accomplished a theoretically endless supply of the reagent would be available and inexpensively produced.

Another recent technology which may be applicable to testing of indoor environments for pathogenic amoebae is the use of gene probes [32]. Several techniques have been used in developing gene probes. Probes have been developed from ribosomal RNA and mitochondrial DNA (*Leishmania mexicana* and *L. braziliensis*). These latter tests are being adapted for field use to identify the organisms not only in human blood smears, but in infected insects as well. While not currently available for free-living amoebae, they could be prepared if warranted. Preparation of DNA probes for protozoa of interest, like monoclonal antibodies, is initially labor intensive, but once prepared can be relatively easily replicated and stored for future use.

When tools such as the monoclonal antibodies and DNA probes are available, more rapid and specific analysis of indoor environments for suspect microbial pathogens, allergens or both can be made. For instance, in the previously discussed outbreak of humidifier fever associated with the presence of *Naegleria* and *Naegleria* antigens which took months of laboratory work to define, the newer techniques may be able to detect these agents or their antigens on site or within days.

Amoebae collected in impinger air samples taken at such a site could be concentrated by centrifugation and subjected to flow cytometry analysis or direct fluorescent antibody analysis using monoclonal antibodies generic for *Naegleria* or *Acanthamoebae* or specific for individual pathogenic species. Alternatively, DNA probe analysis of these amoebic harvests could also detect these genera or the specific pathogens. In either case the analysis could be performed in days rather than weeks as presently required by current techniques. We have already shown the efficacy of using monoclonal or polyclonal antibodies in conjunction with flow cytometry to rapidly detect pathogenic *Naegleria* in environmental samples [33].

In cooling towers and hot water heaters where amplification of *Legionella* by free-living amoebae is a proven problem, monoclonal antibodies could be used to detect these partnerships. Monoclonal antibodies made against both the amoebae and *Legionella* and tagged with fluorescent dyes of different wavelengths could be used to react against the microbial partners. By using flow cytometry, the population of amoebae and *Legionella* not in partnership, as well as the population of amoebae/*Legionella* consortia, could be determined and even separated by cell sorting devices attached to the flow cytometer. Knowing the degree, if any, of amoebae/*Legionella* interaction may be important in proceeding to remedial action since bacteria associated with free-living amoebae have been shown to be more resistant to biocidal action [34].

For detection of amoebic or other microbial antigens in humidifier water or other potential sources of such allergenic material, monoclonal antibodies could also be attached to fiber optic probes. By immersing these probes into potential source fluids and exciting the probes with an appropriate wavelength laser, antigen reacting with the attached monoclonal antibody, as indicated by the spectral response, may be detected. This technology has already been demonstrated for benzopyrene and is capable of detecting molecules in nanomolar concentrations [35]. When miniaturized this instrument could be used at test sites for instantaneous detection of amoebic or other potentially allergenic molecules.

Remedial Action

Before considering the question of remedial action, a reiteration of the medical implications of free-living amoebae in the indoor environment is in order.

The industrial hygienist should be aware that such microorganisms can cause severe eye or wound infections which are highly resistant to antibiotic therapy and which can culminate in fatal encephalitis. Also, while not easily aerosolized because of their size, they can support the growth of *Legionella* which is readily aerosolized and is a pulmonary pathogen. If syndromes associated with free-living amoebae appear in the work force and do not respond to antibiotic treatment traditionally effective with analogous bacterial infections, the presence of pathogenic amoebae should be considered and tests undertaken to determine their role in the infection. If shown to be the etiologic agent, further tests should be undertaken to determine if the source of infection is the workplace. These tests should be carried out by laboratories with proven ability and experience in assaying for these types of pathogens. Other than overt wound or eye infection or resultant encephalitis, the industrial hygienist should be aware that some allergic reactions such as "humidifier fever" have been attributed to amoebic antigens. Thus, if tests fail to implicate bacterial or fungal antigens in the workplace, allergic sensitivity to amoebic antigens should be considered. Such sensitivity could be determined by testing the reactivity of patient sera to amoebic extracts by gel diffusion, radioimmunoassay or other serologic tests using the services of qualified laboratories. Also, if *Legionella* infections occur, the role of free-living amoebae in the *Legionella* presence should also be considered. Again, appropriate laboratories can test for the presence of amoebae/*Legionella* consortia.

Only after free-living amoebae have been proven as the etiologic agent of disease or allergy and their source has been shown to be the workplace, should remedial action be undertaken. Decontamination of the sources should include removal of biofilm and scale possibly using a commercial descaler. This should then be followed by treatment with an oxidizing biocide such as sodium hypochlorite or hydrogen peroxide. The latter, while somewhat slower acting, may be preferable because it is less corrosive and safer to use.

The newer technologies we have discussed, while still in their infancy, will greatly enhance our ability to more rapidly detect and isolate protozoan and other pathogens. They will also greatly reduce the time required to implicate specific microorganisms as the causative agents of indoor air health problems so that remedial action can ensue. Miniaturization of the instruments used in these techniques will also allow for transfer of the methodologies from the laboratory to the test site.

REFERENCES

[1] Martinez, A. J., Free-Living Amebas: Natural History, Prevention, Diagnosis, Pathology, and Treatment of Disease, CRC Press, Inc., Boca Raton, Florida, 1985.

[2] Rondanelli, E. G., Ed., Infectious Diseases: Color Atlas Monographs. 1. Amphizoic Amoebae Human Pathology, Piccin, Italy, 1987.

[3] Page, F. C., "A Light- and Electron-Microscopical Study of *Protacanthamoeba caledonica* n. sp., Type-Species of *Protacanthamoeba* n.g. (Amoebida, Acanthamoebidae)," Journal of Protozoology, Vol. 28, 1981, pp. 70-78.

[4] Bogler, S. A., Zarley, C. D., Burianek, L. L., Fuerst, P. A., and Byers, T. J., "Interstrain Mitochondrial DNA Polymorphism Detected in *Acanthamoeba* by Restriction Endonuclease Analysis," Molecular Biochemistry Parasitology, Vol. 8, 1983, pp. 145-163.

[5] De Jonckheere, J., F., "Isoenzyme and Total Protein Analysis by Agarose Isoelectric Focusing, and Taxonomy of the Genus *Acanthamoeba*," Journal of Protozoology, Vol. 30, 1983, pp. 701-706.

[6] Pernin, P., "Isoenzyme Patterns of Pathogenic and Nonpathogenic Thermophilic *Naegleria* Strains by Isoelectric Focusing," International Journal of Parasitology, Vol. 14, 1984, pp. 459-465.

[7] Byers, T. J., Bogler, S. A., and Buriabrek, L. L., "Analysis of Mitochondrial DNA Variation as an Approach to Systematic Relationship in the Genus *Acanthamoeba*," Journal of Protozoology, Vol. 30, 1983, p. 198.

[8] Willaert, E., "Etude Immunoataxonomique des Genres *Naegleria et Acanthamoeba* (Protozoa: Amoebida)," Acta Zoology Pathology Antwerp, Vol. 65, 1976, pp. 1-239.

[9] Martinez, A. J., "Free-Living Amoebae: Pathogenic Aspects. A Review," Protozoology Abstract, Vol. 7, 1983, pp. 293-306.

[10] Edwards, J. H., Griffiths, A. J., Mullins, J., "Protozoa as Sources of Antigen in 'Humidifier Fever'," Nature, Vol. 264, 1976, pp. 438-439.
[11] "Humidifier Fever," Medical Research Council Symposium, Thorax, Vol. 32, 1977, pp. 653-663.
[12] Martinez, A. J., Sotelo-Avila, C., Garcia-Tamayo, J., Takano, M. J., Willaert, E., and Stamm, W. P., "Meningoencephalitis Due to *Acanthamoeba* sp. Pathogenesis and Clinicopathological Study, Acta Neuropathology (Berl), Vol. 37, 1977, pp. 183-191.
[13] Hamburg, A., and De Jonckheere, J. F., "Amoebic Keratitis," Ophthalmologica (Basel), Vol. 181, 1980, pp. 74-80.
[14] Key, S. N., Green, W. R., Willaert, E., and Stevens, A., "Keratitis due to *Acanthamoeba castellani*. A Clinicopathological Case Report," Archives of Ophthalmology, Vol. 98, 1980, pp. 475-479.
[15] Tyndall, R. L. and Dominque, E. L., "Cocultivatiion of Legionella *pneumophila* and Free-living Amoebae," Applied Environmental Microbiology, Vol. 44, 1982, pp. 954-959.
[16] Fields, B. S., Sanden, G. N., Barbaree, J. M., Morrill, W. E., Wadowsky, R. M., White, E. H., and Feeley, J. C., "Intracellular Multiplication of *Legionella pneumophila* in Amoebaae Isolated from Hospital Hot Water Tanks," Current Microbiology, Vol. 18, 1989, pp. 131-137.
[17] Griffin, J. L., "Temperature Tolerance of Pathogenic and Nonpathogenic Free-Living Amoebas," Science, Vol. 178, 1972, p. 869.
[18] Tyndall, R. L., "Environmental Isolation of Pathogenic *Naegleria*," CRC Critical Reviews in Environmental Control, Vol. 13, No. 3, 1984, pp. 195-226.
[19] Duma, R. J., "Study of Pathogenic Free-Living Amebas in Fresh-Water Lakes in Virginia," EPA-600/S1-80-037, February 1981.
[20] Wellings, F. M., Amuso, P. T., Lewis, A. L., Farmelo, M. J., Moody, D. J., and Osikowicz, C. L., "Pathogenic *Naegleria*: Distribution in Nature," U.S. Environmental Protection Agency-600-/1-79-018, 1979.
[21] Brown, T. J., Cursons, R. T. M., Keys, E. A., Marks, M., and Miles, M., "The Occurrence and Distribution of Pathogenic Free-Living Amoebae in Thermal Areas of the North Island of New Zealand," New Zealand Journal of Marsh Freshwater Research, Vol. 17, 1983, p. 59.
[22] Kadlec, V., Skvarova, J., Cerva, L., and Nebazniva, D., "Virulent *Naegleria fowleri* in Indoor Swimming Pool," Folia Parasitology (Praha), Vol. 27, 1980, p. 11.
[23] Hecht, R. H., Cohen, A., Stoner, J., and Irwin, C., "Primary Amebic Meningoencephalitis in California," California Medicine, Vol. 117, 1972, p. 69.
[24] Kyle, D. E., and Noblet, G. P., "Vertical Distribution of Potentially Pathogenic Free-Living Amoebae in Freshwater Lakes," Journal of Protozoology, Vol. 33, 1985, pp. 422-434.
[25] Fliermans, C. B., Tyndall, R. L., Dominque, E. L., and Willaert, E. J. P., "Isolation of *Naegleria fowleri* from Artificially Heated Waters," Journal of Thermal Biology, Vol. 4, 1979, pp. 303-305.

[26] Lawande, R. V., Abraham, S. N., Jojn, I., and Egler, L. J., "Recovery of Soil Amebas from the Nasal Passages of Children During the Dusty Harmattan Period in Zaria," American Journal of Clinical Pathology, Vol. 71, 1979, pp. 201-203.
[27] Tyndall, R. L., Ironside, K. S., Metler, P. L., Tan, E. L., Hazen, T. C., and Fliermans, C. B., "Effect of Thermal Additions on the Density and Distribution of Thermophilic Amoebae and Pathogenic *Naegleria fowleri* in a Newly Created Cooling Lake," Applied and Environmental Microbiology, Vol. 55, No. 3, March 1989, pp. 722-732.
[28] Lioy, P. J., and Lioy, M. J. Y., Eds., Air Sampling Instruments for Evaluation of Atmospheric Contaminants, American Conference of Governmental Industrial Hygienists, Cincinnati, Ohio, 6th edition, 1983.
[29] Edwards, J. H., "Microbial and Immunological Investigations and Remedial Action after an Outbreak of Humidifier Fever," British Journal of Industrial Medicine, Vol. 37, 1980, pp. 55-62.
[30] Cerva, L., "Studies of *Limax amoebae* in a Swimming Pool," Hydrobiologia, Vol. 38, 1971, p. 141.
[31] Haynes, B. F., and Eisenbarth, G. S., Ed., Monoclonal Antibodies: Probes for the Study of Autoimmunity and Immunodeficiency, Academic Press, Inc. (London), 1983.
[32] Lerman, L. S., Ed., DNA Probes: Applications in Genetic and Infectious Disease and Cancer, Current Communications in Molecular Biology, Cold Spring Harbor Laboratory, 1986.
[33] Muldrow, L. L., Tyndall, R. L., and Fliermans, C. B., "Application of Flow Cytometry to Studies of Pathogenic Free-Living Amoebae," Applied Environmental Microbiology, Vol. 44, 1982, p. 1258.
[34] King, C. H., Shotts, E. B., Jr., Wooley, R. E., and Porter, K. G., "Survival of Coliforms and Bacterial Pathogens within Protozoa during Chlorination," Applied Environmental Microbiology, Vol. 54, 1988, pp. 3023-3033.
[35] Tromberg, B. J., Sepaniak, M. J., Alarie, J. P., Vo-Dinh, T., and Santella, R. M., "Development of Antibody-based Fiber-optic Sensors for Detection of a Benzo[a]pyrene Metabolite," Analytical Chemistry, Vol. 60, 1988, pp. 1901-1908.

DISCUSSION

How can amoebae be controlled in hot tubs and other aquatic systems?

CLOSURE

Frequent cleaning to remove biofilm and treatment with oxidizing biocides is an effective way to control amoebae.

DISCUSSION

Are *Legionella* in amoebae vacuoles protected from chlorine?

CLOSURE

Legionella and other bacteria have been found to be up to fifty times more resistant to chlorine when sequestered within protozoa.

DISCUSSION

Have amoebae been found in bottled water or coffee machines?

CLOSURE

Amoebae have been found occasionally in bottled and tap water and so could be found in coffee makers. However, they were not pathogenic and posed no obvious problem.

DISCUSSION
How long would bacteria sequestered within amoebae remain viable after being aerosolized?
CLOSURE
To my knowledge no studies have addressed this question. Due to their size, amoebae would not be expected to travel far from the point of aerosolization and survival of the associated bacteria would depend in part on the milieu in which the amoebae was deposited.
DISCUSSION
Have studies shown an association between the presence of *Legionella* and free-living amoebae in environmental samples?
CLOSURE
Preliminary studies of *Legionella* in hot water heaters indicates their presence may depend on an association with amoebae.
DISCUSSION
If you suspect that amoebae from a humidifier were causing "humidifier fever" what would be the best way to test for the suspect etiologic agent?
CLOSURE
Analysis of bulk samples from the reservoir would determine which amoebae were present. Serologic tests would indicate whether the patient was reactive to any of the isolates. Since intact amoebae would not be easily aerosolized, reaction of the patient would be more likely with airborne antigens rather than intact amoebae.
DISCUSSION
Are there plans to use monoclonal antibodies to detect amoebae?
CLOSURE
We are presently using monoclonal antibodies to detect pathogenic *Naegleria* in heated waters.
DISCUSSION
Can *Acanthamoebae* incorporate bacteria other than *Legionella*?
CLOSURE
In laboratory experiments *Acanthamoebae* have shown the ability to harbor a variety of gram-negative bacteria and by doing so make them less amenable to destruction by chlorine.
DISCUSSION
Are protozoa killed by infection with *Legionella*?
CLOSURE
There is some evidence from laboratory studies indicating that *Legionella* infected amoebae cultures die more rapidly than their uninfected counterparts.
DISCUSSION
What temperature would eliminate amoebae in hot water tanks?
CLOSURE
A temperature of 60°C would greatly reduce the amoebae population.
DISCUSSION
Could amoebae amplify the presence of bacterial endotoxin?
CLOSURE
Endotoxin originating from those bacteria whose growth is supported by amoebae would be indirectly amplified by the amoebae. The destructive effect of amoebae on some bacteria could also release the associated endotoxin and make it more aerosolizable.

Joseph F. Plouffe

NEW MICROORGANISMS AND THEIR HEALTH RISK

REFERENCE: Plouffe, J.F., "New Microorganisms and Their Health Risk," Biological Contaminants in Indoor Environments, ASTM STP 1071, Philip R. Morey, James C. Feeley, Sr., James A. Otten, Editors, American Society for Testing and Materials, Philadelphia, 1990.

ABSTRACT: A vast array of microorganisms have been associated with pulmonary infections. Our knowledge is still incomplete. There are still unrecognized pulmonary pathogens. Future research should concentrate on new culture media, immunocompromised patients, epidemics and environmental isolates.

KEYWORDS: Pneumonia, pulmonary pathogens, environmental organisms

Introduction

My task this evening is to speculate on the topic of new microorganisms and their health risks as related to respiratory infections. I will try not to repeat information which you have heard over the last two days. You have already been presented with a broad overview of microbiology: gram negative organisms including *Legionella sp.*, gram positive organisms, rickettsia, mycobacteria, viruses, chlamydia, fungi and protozoa. The impressive array of lectures on these microorganisms would seem to cover the extent of our knowledge of interactions between microorganisms and the human host.

I could generate a hypothesis that states that now in 1989, we have learned about all potential organisms which can cause respiratory problems. This would make my task much easier. Let us examine the validity of this hypothesis. One approach would be to look back historically to see what people thought about the causes of pneumonia at various times. Organisms will be referred to as they were cited in the past literature. This paper is not meant to be an

Dr. Joseph F. Plouffe is a Professor of Internal Medicine, Division of Infectious Diseases at The Ohio State University Hospitals, 410 West 10th Avenue, Columbus, Ohio 43210-1228.

exhaustive review of pneumonia. Only selected papers will be cited to emphasize certain points.

Historical Background

Going back to the time of the ancient Greeks, the knowledge regarding respiratory infections was mainly confined to clinical descriptions of patients with pneumonia. It has only been in the last one hundred years that specific organisms were associated with disease.

In the early 1900's, cases of pneumonia were tabulated. There was a tendency towards descriptive terminology in these series. Cases were either described as lobar which was typically caused by the *Streptococcus pneumoniae,* pneumococcus, or bronchopneumonia, which could also be caused by the pneumococcus, as well as other etiologies. In Osler's 1920 textbook of medicine published the year after his death from pneumonia, he reported there were approximately 100,000 deaths due to pneumonia each year, with a rate of 200 per 100,000 population. Approximately 60% were lobar pneumonia, 33% were bronchopneumonia and the rest were unclassified. The most common isolate during these times was the pneumococcus.[1]

In the late 1920's and early 1930's, another series of pneumonia cases was described from Harlem Hospital by Dr. Bullowa.[2] They had 3,000 cases of pneumonia of which 84% were felt to be caused by pneumococcus, 9% by *S. pyogenes* and small percentages due to *Staphylococcus aureus, Klebsiella pneumoniae* and a variety of miscellaneous organisms including *Mycobacterium tuberculosis*, *Salmonella sp.*, *Haemophilus sp., Pasteurella tularensis*, *Corynebacterium diphtheriae*, Influenza virus, Rubeola and *Chlamydia psittaci*. Only 2% of their cases did not have an etiology. During this time, new descriptive terminology was prevalent in the literature; typical pneumonia versus atypical pneumonia. These terms correlated reasonably well with the chest x-ray, clinical and pathologic descriptions of lobar and bronchopneumonia.[3]

In the 1940's mycoplasma was first cultured by Eaton and found to be important in the cause of pulmonary infections.[4] In a 1953 textbook of medicine edited by Garland and Phillips, only a passing mention was made of an undefined pneumotropic virus responsive to tetracycline.[5] One of the problems with categorizing series of pneumonias, is that many times series only included patients that were hospitalized. Many patients with atypical pneumonias did not require hospitalization and were not included in many of these early series.

In the late 1960's there was a series of cases of pneumonia reported from Hannaman Medical College of Philadelphia: only 54% of the cases of pneumonia were caused by *S. pneumononiae*, 6% were caused by *K. pneumoniae*, 2% by *S. aureus*, 5% by Herpes simplex, 3% by *Enterobacter sp.*, 0.5% by Adenovirus. There was a large category, 30%, of undefined pneumonias.[6] This series recognized that many patients fell into an undiagnosed category. In the late 1960's and early 1970's we also started to see several other changes in the spectrum of organisms causing pneumonia. The gram negative organisms which had previously been mainly *H. influenzae* and *K. pneumoniae*, now expanded to multiple gram negative organisms including *Pseudomonas aeruginosa, Proteus sp., Serratia sp.* and *Escherichia coli.* These organisms were frequently associated with nosocomial pneumonia; patients acquiring their pneumonia in the hospital.[7,8]

Another group of organisms which started to gain recognition during this time, were the anaerobic organisms. Excellent studies were performed by Dr. Feingold and investigators at the Wadsworth VA Hospital in Los Angeles which demonstrated the presence of anaerobes in aspiration pneumonia.[9] They characterized the mixed etiology of aspiration pneumonia by transtracheal aspiration by which they were able to bypass the heavily colonized oropharyngeal cavity and obtain anaerobic cultures from the lower respiratory tract.

In the 1970's, the role of fungi causing respiratory disease was better understood. Much of this had to do with an increasing population of immunocompromised patients who developed fungal infections of their lungs, as well as several large outbreaks.[10,11]

In the late 1970's, the next group of additions to etiologic agents for pneumonia was *Legionella*.[12,13] Dr. Dennis has already described this new family of organisms which exists in water and causes respiratory disease in susceptible hosts.

During this decade, the gram negative pneumonias increased in frequency, especially in the immunocompromised and neutropenic patients. A major factor here is the advent of newer and broader antimicrobials which eradicated most of the normal oropharyngeal organisms and allowed for colonization with these resistant gram negative bacilli.

Another bacterial organism *Moraxella catarrhalis* (formerly *Branhamella catarrhalis)*, which previously had been named *Neisseria catarrhalis* and felt to be a commensual organism in the oropharynx, was demonstrated to cause pneumonia in patients with chronic obstructive pulmonary disease.[14]

In the current decade, another new agent, the TWAR strain of chlamydia, called *Chlamydia pneumoniae* was recognized as a cause of atypical pneumonia.[15]

Also, a major factor in the 1980's has been an increasing prevalence of patients with acquired immunodeficiency syndrome. These patients are at high risk of developing pulmonary infections, mainly due to *Pneumocystis carinii*, but also due to *Mycobacterium tuberculosis* as well as atypical mycobacteria.[16,17] These patients are also susceptible to bacterial infections with *S. pneumoniae*, *Salmonella sp.* and a gram positive bacilli *Rhodococcus equii*. The AIDS patients also have problems with other organisms which require intact cellular immunity. Specific examples include *Histoplasma capsulatum* and *Cryptococcus neoformans*.[18,19]

In review, in the 1940's, all pneumonia was felt to be due to pneumococcus. In each subsequent decade, another major group of organisms was recognized as a cause of respiratory infection. In the 1950's, *Mycoplasma pneumoniae*, in the 1960's gram negative organisms and anaerobes, in the 1970's *Legionella pneumophila* and *Moraxella catarrhalis*, in the 1980's *Chlamydia pneumoniae*, as well as an increasing prevalence of many opportunistic organisms in the AIDS population.

Were all these organisms present and causing pneumonia in the 1940's and just not recognized? What changes occurred with *Streptococcus pyogenes*? Previously a cause of pneumonia, it is rarely associated with pneumonia in current times. There have been changes in our patient population since the 1940's that allow for some of these other organisms to infect this new susceptible patient population. There have been environmental changes which have allowed some organisms to proliferate and then attack susceptible populations.

If we look at the development of antibiotics over the last 40 years, after the release of each class of antibiotics, resistant strains will develop over the next 5 to 10 years. During the last 40 years, we have made significant strides in controlling many previously fatal diseases such as leukemia, malignancies and renal failure. The populations of patients who survive these diseases are obviously different from the patients in the 1940's. Frequently, because of their therapy, they are susceptible to organisms which previously had not caused problems. The elderly population has increased dramatically over the past 100 years. These patients may be susceptible to organisms that younger individuals are not.

An important concept is that of pathogenic or virulent types of organisms versus those that are felt to be avirulent. The concept of virulence is a relative phenomenon. Many organisms present in the environment, if concentrated to a high enough degree and presented to a susceptible patient, will be able to cause illness.

Two more recent series of pneumonia demonstrate the inability to make an etiologic diagnosis. The first, published by White et al from Bristol, UK, looked at pneumonia patients admitted to the hospital between 1974 and 1980.[20] There were 210 cases: 14% due to Influenza virus, 14% due to *M. pneumoniae*, 11% due to *S. pneumoniae*, 4% due to *S. aureus*, 2% due to *H. influenzae* and 50% had no pathogen isolated. The second series from Nottingham, UK by Woodhead et al looked at 236 patients presenting to community physicians with pneumonia from 1984 to 1985.[21] Patients were evaluated whether or not they required hospitalization. In this series, 36% were due to *S. pneumoniae*, 10% due to *H. influenzae*, 6% due to Influenza A. Organisms which represented 1-2% each, included: *S. aureus, E. coli, L. pneumophila, M. tuberculosis, M. pneumoniae, C. psittaci*, Influenza B, and other viruses. No pathogen was isolated or identified by serology in 45% of their patients. These two studies suggest that there are organisms present in the environment which are causing pneumonia which we have not yet defined.

New Etiological Agents: Past

I think it would be beneficial to look back at the situations which led to the discovery of new pneumonia agents. We can see if there were common threads among the investigations which led to the description of these new etiologic agents.

Mycoplasma pneumoniae. It was known in the early 1900's that a pneumonia was caused in cattle by Pleuropneumonia organisms (PPO). In the 1930's, astute observations by clinicians described an atypical pneumonic illness which was different from the typical pneumococcal clinical presentation. In the 1940's, the Eaton agent was shown to infect animals [4] and a serologic response (cold agglutinin) was seen in some patients with atypical pneumonia. Another decade later, a fluorescent antibody stain was developed by Liu which was used in experimental models to show that this organism caused a pulmonary infection.[22] It was recognized that this organism was similar to the Pleuropneumonia organisms from cattle. In the 1960's, Chanock was able to culture mycoplasma on cellfree medium and show that tetracycline therapy was beneficial.[23]

Legionella. In 1947 the initial isolate of *Legionella* was cultured from a Navy scuba diver after guinea pig innoculation.[24] At the time, since no serologic response was demonstrated in the patient, the organism was felt to be a contaminant. It wasn't until 1976 when the large epidemic occurred in Philadelphia, that the eventual cause of Legionnaires disease was determined.[12] Even with this large outbreak, it took approximately six months for the organisms to be demonstrated.[13] Eventually, *Legionella* were grown on cellfree medium and serologic tests were developed. Once these serologic tests were available, outbreaks which had occurred prior to 1976 were also demonstrated to be caused by *Legionella* by study of sera that was saved.[25] Through the 1980's, further epidemiologic studies had defined a complex picture of *Legionella* infections occurring as hospital acquired infections, as well as in small community outbreaks and in some geographic areas where *Legionella* seemed to be endemic.

Moraxella catarrhalis. This organism has been known for a long time but felt to be a commensual organism in the oropharynx. In the late 1970's, transtracheal aspirates show that this organism was present in pure culture in 5-10% of patients with chronic obstructive pulmonary disease and pneumonia.[14] In the 1980's, a serologic test was developed and shown that patients with pneumonia seroconverted to *M. catarrhalis.*

Chlamydia pneumoniae. This organism was initially isolated in 1965 during a trachoma survey.[19] It was recognized that this organism was antigenically different from *Chlamydia trachoma.* In the 1970's a microimmunofluorescence antibody test was developed.[26] In 1978, a fortuitous outbreak of pneumonia was found in Finland. Investigators were doing tuberculosis screening chest x-rays and found relatively asymptomatic young adults with abnormal chest x-rays. Four years later, it was demonstrated that this organism (TWAR strain of chlamydia) caused serologic conversion in these patients.[19] In the past several years, prospective studies done at the University of Washington Health Center, have shown that young patients with atypical pneumonias seroconverted to *Chlamydia pneumoniae*. These investigators were also able to culture the *Chlamydia pneumoniae* from the pharynx of these patients.[26] Further serologic surveys in large populations have shown that significant percentages of the population have antibodies to *Chlamydia pneumoniae*.[27,28]

What are the common threads that led to the discoveries of each of these organisms as causes of pulmonary infection?

1. These organisms had previously been identified but not necessarily considered to be human pathogens.
2. Most did not grow on standard media used to screen for respiratory pathogens.
3. Either astute clinical observations or an epidemic situation occurred.
4. Technical advances were made either in the development of serologic tests or media.
5. Epidemiologic studies were done that showed the extent of the disease caused by these organisms.

<u>New Etiological Agents: Future</u>

What factors would be helpful in identifying new pneumonia agents?

1. The microbe must be able to be cultured either from the environment or from respiratory secretions.
2. The agent must be visualized by stain or be able to be characterized morphologically.
3. A link must be made between the organism and a clinical disease using epidemiologic studies, cultures, stains, serologic studies, and antigen detection methodology.
4. Methods should be developed to screen environmental organisms and their potential to cause disease.

There are a vast number of undefined microorganisms present in the environment. It would be a large task to look at each of these organisms to determine the potential for causing respiratory disease. One way to approach this problem would be to look very carefully at the undiagnosed cases of pneumonia. To understand which organisms would escape our current diagnostic techniques, it is necessary to know how a respiratory specimen is evaluated. Most clinical microbiology laboratories perform gram stains on expectorated sputum. There has always been a significant debate in the medical literature as to the usefulness of this technique. It may, however, be very important in defining a particular specimen which has a predominant organism and negative culture. Also, most clinical microbiology laboratories culture respiratory specimens on limited media. The media that is used has been defined to grow "pathogens." Since approximately 40-50% of patients with pneumonia do not grow a pathogen, perhaps some adjustments in how respiratory secretions are handled by cultural techniques are necessary.

In a paper presented at the latest American Society of Microbiology Meeting, concerning flora present in whirlpools and sauna, approximately 50% of the gram negative organisms isolated were not able to be identified by investigators from Centers for Disease Control.[29] These organisms fall into the large group of heterotrophic gram negative organisms present in our water supply.

Several years ago we did a survey of multiple hospitals looking a colonization with Legionella. During the survey, we also cultured for heterotrophs, using the R2A medium defined by Geldrich and Reisner.[30] Several of the hospitals had extremely large numbers of heterotrophs in portable water systems. Typically, these organisms will not grow on standard media used in evaluation of a respiratory specimen. There is some work currently being done by Al Dufour and his coworkers at the EPA in Cincinnati, examining water organisms for potential virulence factors. These investigators have also described an animal model which can be used to test for the ability of these organisms to cause respiratory disease.[31] I believe that these investigations should be continued as a high priority item. This large group of water organisms has never been well categorized. This is mainly because they have not been associated with human disease and have generated little interest. With the current advances in technology, it should be possible to classify these organisms.

Thorough studies should be instituted on patients with pneumonia of undefined etiology. Respiratory secretions should be cultured on special media designed to grow unusual organisms. Serologic tests can be developed which will allow us to determine if a patient has had an immune response to these water organisms.

Another potential place where we may be able to make some inroads in finding unusual causes of pneumonia, is the AIDS population. These patients are susceptible to respiratory infections. Predominantly, they develop infections with *Pneumocystis carinii*. However, they are also prone to develop infections due to bacterial organisms such as pneumococcus. Since many of these patients undergo a bronchoscopy with bronchoalveolar lavage, very good specimens can be obtained from the lower respiratory tree for complete evaluation. The current methodology for evaluating a bronchoalveolar lavage specimen include culturing for routine bacteria, fungi, acidfast bacilli (AFB), *Legionella*, and viruses. Special stains are done to look for pneumocystis, fungi, *Legionella* and AFB. These methods may vary from one clinical laboratory to another.

A recent study from the National Institutes of Health reviewed 152 episodes of clinical pneumonia in 110 patients with AIDS.[32] These patients had a very thorough workup with bronchoalveolar lavage and transbronchial biopsies in 102 episodes, bronchoalveolar lavage alone in 27, open lung biopsies in 23 patients. Overall, 41% of the patients had *P. carinii* as the cause of their pneumonitis. Kaposi sarcoma was found in 6% of the patients. Twelve percent had non-diagnostic studies. Most of these patients only had bronchoalveolar lavage. Cytomegalovirus was seen in 6% of the patients, bacterial infections in 4%, and *Mycobacterium sp.* in 3% and 1% with *C. neoformans*. There was a large group of 27% or 41 patients who had a non-specific interstitial pneumonitis. Attempts to isolate or identify organisms in these patients were not successful.

Since case reports of unusual causes of pneumonia have sometimes been an initial clue, this patient population with this predilection to respiratory infections may serve to show us which environmental organisms can cause disease.

Another interesting facet of respiratory infections has been the realization that colonization of the upper gastrointestinal (UGI) tract may be important as a predisposing factor to the development of respiratory infections. This is a relatively new phenomenon due to the development of new medications such as the H2 blockers which will reduce gastric acid secretion. Also, as our population ages, more people without gastric acid will be present. Gastric acid serves as a very potent natural barrier to colonization of the upper GI tract with microorganisms. Several studies have shown that patients treated with H2 blockers had higher pH's, higher concentrations of gram negative bacilli and higher rates of nosocomial pneumonia, than did the control group with normal gastric acid.[33,34,35] In at least one outbreak of Legionnaires disease investigated by CDC, the taking of antacids was an independent risk factor for development of disease. This was independent of steroid use. Is it possible that the UGI tract is colonized with *Legionella*? Another interesting association is that of development of tuberculosis in patients with gastrectomies. With our current knowledge of the pathophysiology of *Mycobacterium tuberculosis* infection, it is difficult to understand this concept. I think further studies looking at colonization of the UGI tract are important. It would be important to look for organisms other than the typical pathogens in gastric secretions. Investigators from the Netherlands studied prevention of nosocomial pneumonia in trauma patients.[36] They used oropharyngeal and intestinal decontamination with an ointment of antibiotics and found that secondary colonization and infection of the respiratory tract with gram negative bacilli was significantly reduced when compared to controls.

HYPOTHETICAL OUTBREAK

On July 9, 1989 an eighty-four year old gentleman presented to Boulder General Hospital with fever, confusion, cough, and shortness of breath. Initial evaluation of the patient revealed that he had a pneumonia. Sputum cultures grew normal flora. The patient was initially treated with a broad spectrum cephalosporin with no response. On the following day, three other people, all in their early eighties were all admitted to the same hospital with similar illnesses. Initial review revealed that all four patients lived in the same senior citizens' complex. These patients were treated with erythromycin on the presumption that an outbreak of Legionnaires disease was occurring. The second group of patients started to respond to antibiotic therapy. Cultures of their sputum also grew no pathogenic organisms. Direct fluorescent antibody stains looking for *Legionella* were negative. Patients in general were in reasonably good health for their advanced age. Two patients had diabetes mellitus controlled with oral agents and two patients had peptic ulcer disease and were treated with H2 blockers. Further investigation into residents of this senior citizens' home revealed that one other patient had a similar illness and had been hospitalized in Denver. Interviews with the patients and other members of the senior citizens' home were performed. An unusual sequence of events was discovered. On July 4, the subjects all participated in a wine and cheese fourth of July celebration. The celebration was somewhat dampened by the fact that one of their close friends had died of a heart attack in the preceding week. After consuming several glasses of wine, one of the more active members of the group, decided that eating a lot of cheese would be bad for their coronary arteries. He proceeded to take the various cheeses which were present at the party and dump them down the garbage disposal. There was some disagreement as to this action and several people were involved in heated discussion in the kitchen as the cheese was being disposed.

Investigators from the local health department sampled the water supplies from the senior citizens home as well as their spa for *Legionella* and were unable to grow *Legionella*. Because of the unusual epidemiologic story, several of the sputum specimens were cultured using selective media and enrichment techniques. Specifically, cold enrichment techniques were done looking for *Listeria monocytogenes*. After several weeks, sputum from two of the patients grew *Listeria monocytogenes*.

Is *Listeria monocytogenes* a potential etiologic agent in pneumonia? It certainly fulfills the criteria that we have listed of potential pneumonia agents. It has been isolated previously. It causes disease in humans, but not necessarily known to cause a significant number of respiratory infections. *Listeria* can cause pneumonia in mice.[37] The organism does not grow readily on standard media if other flora are present. The organism would probably be ignored in stains of expectorated sputum thinking that it was typical oropharyngeal flora. The organism is known to cause epidemics.[38] Several large outbreaks associated with cheese, unpasteurized milk, and cole slaw have been described. The major problem is in pregnant females where there is significant morbidity and mortality to the fetus.[38] There can also be significant mortality in immunocompromised patients.[39] In sporadic cases of listeriosis, the attack rates are highest among newborns and persons over the age of 70 with an incidence of approximately 11 per million population. *Listeria* has been detected in 2-3% of processed dairy products tested by the FDA.[38]

It was presumed that at least one of the cheeses purchased for the party was contaminated with *Listeria*. By putting the cheese through the garbage disposal, it was hypothesized that an aerosol was generated. The elderly patients exposed to this aerosol then became infected with this organism. Fortunately, the organism is somewhat susceptible to erythromycin. By treating the patients for presumed Legionnaires disease, an active antibiotic was used.

I think this scenario shows how an outbreak can cause the recognition of a pulmonary pathogen.

In conclusion, I think we have sufficient evidence to reject the initial hypothesis. There will be other respiratory pathogens identified. It will be our task to be vigilent and inquisitive in seeking out these new pathogens. A combination of environmental sampling, new cultural techniques and associations of organisms with clinical findings should help us identify these new pathogens.

REFERENCES

[1] Osler W., McCrae, T., eds. The Principles and Practices of Medicine. D. Appleton and Company, New York, 1923.

[2] Bullowa, J.G.M., ed. In the Management of the Pneumonias. Oxford University Press, New York, 1937.

[3] Reimann, H.A. "An acute infection of the respiratory tract with atypical pneumonia. A disease entity probably caused by a filtrable virus." The Journal of the American Medical Association, Vol. 251, No. 7, February 1984, pp. 936-944.

[4] Eaton, M.D., Meiklejohn, G., Van Herick, W. "Studies on the etiology of primary atypical pneumonia. A filterable agent transmissible to cotton rats, hamsters, and chick embryos." Journal of Experimental Medicine, Vol. 79, 1944, pp. 649-667.

[5] Garland, H.G., Phillips, W. (eds). Medicine. Macmillan and Co., Limited, London, 1953.

[6] Reimann, H.A. The Pneumonias. Warren H. Green, Inc., St. Louis, 1971.

[7] LaForce, F.M. "Hospital-acquired gram-negative rod pneumonia: An overview." The American Journal of Medicine, Vol. 79, 1981, pp. 664-669.

[8] Johanson, W.G., Pierce, A.K., Sanford, J.P. "Changing pharyngeal bacterial flora of hospitalized patients: Emergence of gram-negative bacilli." The New England Journal of Medicine, Vol. 281, 1969, pp. 1137-1170.

[9] Bartlett, J.G., Rosenblatt, J.E., Feingold, S.M. "Percutaneous transtracheal aspirates in the diagnosis of anaerobic pulmonary infection." Annals of Internal Medicine, Vol. 70, 1973, pp. 535-540.

[10] Kauffman, C.A., Israel, K.S., Smith, J.W., et al. "Histoplasmosis in Immunosuppressed Patients." The American Journal of Medicine, Vol. 64, June 1978, pp. 923-932.

[11] Wheat, L.J., Slama, T.G., Norton, J.A., et al. "Risk factors for disseminated or fatal histoplasmosis. Analysis of a large urban outbreak." Annals of Internal Medicine, Vol. 96, February 1982, pp. 159-163.

[12] Fraser, D.W., Tsai, T. R., Orenstein, W., et al. "Legionnaires' disease. Description of an epidemic of

pneumonia." The New England Journal of Medicine, Vol. 297, No. 22, December 1977, pp. 1189-1196.

[13] McDade, J.E., Shepard, C.C., Fraser, D.W., et al. "Legionnaires' disease. Isolation of a bacterium and demonstration of its role in other respiratory disease." The New England Journal of Medicine, Vol.297, No. 22, December 1977, pp. 1197-1198.

[14] Ninane, G., Joly, J., Kraytman, M. "Bronchopulmonary infection due to Branhamella catarrhalis: 11 cases assessed by transtracheal puncture." The British Medical Journal, Vol. 1, 1978, pp. 276-278.

[15] Saikku, P., Wang, S.P., Kleemola, M., et al. "An epidemic of mild pneumonias due to an unusual strain of Chlamydia psittaci." The Journal of Infectious Diseases, Vol. 151, 1985, pp. 832-839.

[16] Murray, J.F., Felton, C.P., Garay, S.M., et al. "Pulmonary complications of the acquired immunodeficiency syndrome: reprint of a National Heart, Lung and Blood Institute Workshop." The New England Journal of Medicine, Vol. 310, 1984, pp. 1682-1688.

[17] Stover, D.E., White, D.A., Romano, P.A., Gillene, R.A., Robeson, W.A. "Spectrum of pulmonary diseases associated with the acquired immune deficiency syndrome." The American Journal of Medicine, Vol. 98, 1985, pp. 429-437.

[18] Wheat, L.J., Slama, T.G., Zeckel, M.L. "Histoplasmosis in the acquired immune deficiency syndrome." The American Journal of Medicine, Vol. 78, February 1985, pp. 203-210.

[19] Kovacs, J.A., Kovacs, A.A., Poles M., et al. "Cryptococcus in the acquired immunodeficiency syndrome." Annals of Internal Medicine, Vol. 103, 1985, pp. 533-538.

[20] White, R.J., Blainey, A.D., Harrison, K.J., Clarke, S.K.R. "Causes of pneumonia presenting to a district general hospital." Thorax, Vol. 36, 1981, pp. 566-570.

[21] Woodhead, M.A., MacFarlane, J.T., McCracken, J.S.M., Rose, D.H., Finch, R.G. "Prospective study of the aetiology and outcome of pneumonia in the community." The Lancet, Vol. i, March 1987, pp. 671-674.

[22] Liu, C. "Studies on primary atypical pneumonia; localization, isolation, and cultivation of a virus in chick embryos." Journal of Experimental Medicine, Vol. 104, 1957, pp. 455-466.

[23] Chanock, R.M., Hayflick, L., Burile, M.F. "Growth on artificial medium of an agent associated with atypical pneumonia and its identification as a PPLO." Proceedings from the National Academy of Science, Vol. 48, 1962, pp. 41-49.

[24] Brenner, D.J., Steigerwalt, A.G., McDade, J.E. "Classification of the Legionnaires' disease bacterium, Legionella pneumophila, genus novum, species nova, of the family legionellaceae, family nova. Annals of Internal Medicine, Vol. 90, 1979, pp. 656-658.

[25] Glick, T.H., Gregg, M.B., Buman, B., Mollison, G., Rhodes, W.W., Kassanoff, I. "Pontiac fever. An epidemic of unknown etiology in a health department. 1. Clinical and epidemiologic aspects." The American Journal of Epidemiology, Vol. 107, 1978, pp. 149-160.

[26] Grayston, J.T., Cho-chou, K., San-pin, W., Altman, J. "A new Chlamydia psittaci strain, TWAR, isolated in acute respiratory tract infections." The New England Journal of Medicine, Vol. 315, No. 3, July 1986, pp. 161-168.
[27] Marrie, T.J., Grayston, J.T., San-pin, W., Cho-chou, K. "Pneumonia associated with the TWAR strain of chlamydia." Annals of Internal Medicine, Vol 106, 1987, pp. 507-511.
[28] Kleemola, M., Saikku, P., Visakorpi, R., Wang, S.P., Grayston, J.T. "Epidemics of pneumonia caused by TWAR, a new chlamydia organism, in military trainees in Finland." The Journal of Infectious Diseases, Vol. 157, No. 2, February 1988, pp. 230-236.
[29] McNamara, A.M., Kaylor, B.M., Highsmith, A.K. "Microbial species isolated from high heat, low humidity environments." Abstract Q-98. 89th Annual Meeting of the American Society of Microbiology, 1989.
[30] Reasoner, D.J., Geldreich, E.E. "A new medium for the enumeration and subculture of bacteria from potable water." Applied & Environmental Microbiology, Vol. 49, No. 1, January 1985, pp. 1-7.
[31] Davis-Hoover, W., Dufour, A., Stelena, G., Scarpine, P. "An animal model for studying opportunistic pathogens isolated from drinking water." Abstract Q-120. 89th Annual Meeting of the American Society of Microbiology, 1989.
[32] Suffredini, A.F., Ognibene, F.P., Lack, E.E., Simmons, J.T., et al. "Nonspecific interstitial pneumonitis: a common cause of pulmonary disease in the acquired immunodeficiency syndrome." Annals of Internal Medicine, Vol. 107, No. 1, July 1987, pp. 7-13.
[33] Atherton, S.T., White, D.J. "Stomachs as source of bacteria colonizing respiratory tract during artificial ventilation." The Lancet, Vol. 2, 1982, pp. 908-909.
[34] DuMoulin, G.C., Paterson, D.G., Hedley-Whyte, J., Lisbon, A. "Aspiration of gastric bacteria in antacid-treated patients: a frequent cause of postoperative colonization of the airway." The Lancet, Vol. 1, pp. 242-245.
[35] Driks, M.R., Craven, D.E., Celli, B.R., Manning, M. "Nosocomial pneumonia in intubated patients given sucralfate as compared with antacids or histamine type 2 blockers. The role of gastric colonization." The New England Journal of Medicine, Vol. 317, No. 22, November 1987, pp. 1376-1382.
[36] Stoutenbeek, C.P., Van Saene, H.K.F., Miranda, D.R., et al. "The effect of oropharyngeal decontamination using topical nonabsorbable antibiotics on the incidence of nosocomial respiratory tract infections in multiple trauma patients." Journal of Trauma, Vol. 27, No. 4, April 1987, pp. 357-364.
[37] Lefford, M.J., Amell, L., Warner, S. "Listeria Pneumonitis: Induction of immunity after airborne infection with *Listeria monocytogenes*." Infection & Immunology, Vol. 22, No. 3, December 1978, pp. 746-751.
[38] Gellin, B.G., Broome, C.V. "Listeriosis." The Journal of the American Medical Association, Vol. 261, No. 9, March 1989, pp. 1313-1320.
[39] Stamm, A.M., Dismukes, W.E., Simmons, B.P., et al. "Listeriosis in renal transplant recipients: report of an outbreak and review of 102 cases." Reviews of Infectious

Diseases, Vol. 4, No. 3, May-June 1982, pp. 665-682.

DISCUSSION 1

Won't our new pneumonia etiologic agents generally be antibiotic resistant?

CLOSURE 1

There is no doubt that certain antibiotic resistant organisms will cause pneumonia. One of the examples of this would be the antibiotic resistant Streptococcus pneumoniae. Basically, I feel that these organisms will be the ones with which we are currently familiar. It will require different antibiotic therapy to achieve cure. I think it is likely that we will find other organisms, not currently recognized as pulmonary pathogens which may be susceptible to currently used antibiotics. Many patients with the clinical picture of pneumonia and negative cultures are treated with standard antibiotics and recover. Some of these pneumonias may have responded to these antibiotics, while others may have responded to host defenses.

DISCUSSION 2

Do you think the changing AIDS population (homosexual white males to drug abusers, etc.) may also change the types of pneumonia?

CLOSURE 2

The most common cause of pneumonia in the AIDS population is Pneumocystis carinii. Strategies have been taken to reduce the frequency of this etiology. I suspect we will see other organisms causing respiratory infections in this population, both in the homosexual and drug abuse population. The organisms we may see include members of the mycobacterial family, both typical and atypical.

DISCUSSION 3

Could a "new" pathogen be produced because of selected pressures from mechanical and plumbing systems in new building environments?

CLOSURE 3

I think environmental select pressures will be extremely important in defining new potential pathogens. The prime example of this type of selection occurred with Legionella pneumophila. It is difficult to predict which organisms would be selected by various changes in building environments. One must examine these environments after the changes have been made to survey for potential problem organisms.

DISCUSSION 4

Do "unidentified" bacteria in potable water have any possible health significance?

CLOSURE 4

Most environmental organisms which remain unclassified have not been associated with disease states. However, it is possible that these organisms can cause human disease. If the microbiology laboratories are unable to grow these organisms on their standard mediums, one would not recognize them as a cause of human disease.

DISCUSSION 5

In the 1968 Pontiac Fever outbreak, guinea pig sentinel animals were used. Do you see any future value in detecting new organisms by this technique?

CLOSURE 5

I think that in selected instances, sentinel guinea pigs can provide very useful information. In the Pontiac Fever outbreak, it was only because of these sentinel guinea pigs that the specific organism and its subtype was able to be identified many years later. One potential benefit that the sentinel guinea pigs have over the standard air sampling methodology would be if you had an organism which did not grow on the culture medium being used.

Stephen A. Olenchock

ENDOTOXINS

REFERENCE: Olenchock, S. A., "Endotoxins," Biological Contaminants in Indoor Environments, ASTM STP 1071, Philip R. Morey, James C. Feeley, Sr., James A. Otten, editors, American Society for Testing and Materials, Philadelphia, 1990.

ABSTRACT: Gram-negative bacteria contain endotoxins as integral components of their outer membrane. Endotoxins are potent biological agents, and, as contaminants of dusts, can be associated with acute lung reactivity in exposed individuals. Endotoxins can be quantified in bulk materials, dusts and waters, yet many questions remain to be resolved related to their relative toxicities and other properties.

KEYWORDS: endotoxin, gram-negative bacteria, agriculture, humidifier fever, *Limulus*

CHARACTERISTICS AND SOURCE

Endotoxins are heat stable, lipopolysaccharide-protein complexes that are integral parts of the outer membrane of gram-negative bacteria [1]. Reviews of chemical composition of endotoxins provide exhaustive descriptions of the molecular nature of these entities which will be discussed here only in brief [2, 3, 4, 5].

In the outermost cell membrane, the lipopolysaccharide (LPS) molecule consists of three regions, O-specific popysaccharide (I), core polysaccharide (II), and Lipid A (III). The O-specific polysaccharide chain provides serologic specificity to the different serotypes of gram-negative bacteria. Substitutions, composed of sugars and their sequences on repeating oligosaccharide units, define the antigenic uniqueness for the region I. Markedly less variability is found in the core polysaccharide which can be identical for large groupings of bacteria. A unique ketose, 2-keto-3-desoxyoctonic

Dr. Olenchock is Chief, Immunology Section, Division of Respiratory Disease Studies, National Institute for Occupational Safety and Health, 944 Chestnut Ridge Road, Morgantown, WV 26505 and Adjunct Professor of Microbiology and Immunology, West Virginia University Health Sciences Center, Morgantown, WV 26506.

acid (KDO) can be found in region II. Region III, the least variable chemical region, consists of Lipid A. The predominant endotoxic activities ascribed to endotoxins can be found in Lipid A [2, 3, 4].

Endotoxins are released into the environment after lysis of the bacterial cells and also during active cell growth [6]. In addition, endotoxins from intact bacterial cells can be taken up by macrophages and processed, thus increasing their toxicity [7]. Recently described "cell-bound" endotoxins that occur still attached to bacterial cells have been discussed as important sources of exposure to endotoxins [8].

Gram-negative bacteria and their endotoxins are ubiquitous microorganisms found in the soil, water, and in other living organisms around the world. They are found especially in various materials in agricultural environments, particularly those associated with tasks common to farming [9]. Yet they are not confined to environments that produce large amounts of organic dusts. In addition to environments related to agricultural, gram-negative bacteria and their endotoxins are found as well in environments that are considered traditionally to be non-dusty such as office buildings and libraries where humidification systems are operative [10, 11].

HEALTH EFFECTS

Humoral and cellular host mediation systems are profoundly affected by endotoxins [2, 3, 6]. Effects on the complement and coagulation systems can be observed, as can the direct interactions with many cell types including basophils, mast cells, endothelial cells, macrophages, platelets, polymorphonuclear leukocytes, and T and B lymphocytes [2, 3, 6]. The pulmonary macrophage is the primary target cell for endotoxin-induced damage by inhalation [12], and the human alveolar macrophage in particular is extremely sensitive to the effects of endotoxins _in vitro_ [13]. Additionally, lipopolysaccharide (LPS) has been shown to bind to pulmonary surfactant and alter its surface tension properties which could result in impaired pulmonary function [14].

Systemic signs and symptoms such as chest tightness, cough, shortness of breath, fever, and wheezing that are suggestive of exposure to airborne endotoxins have been reported for workers in sewage treatment plants [15], swine confinement buildings [16, 17], and poultry units [18].

Controlled exposures of human subjects to endotoxin-laden cotton dusts lead to the definition of an association between decreases in acute pulmonary function and the airborne level of endotoxins in the dusts [19, 20, 21]. Cotton mill studies in The People's Republic of China showed also a dose-response relationship between endotoxin levels in cotton dust and chronic lung impairment such as chronic bronchitis among cotton workers [22]. In the acute studies, the thresholds for no pulmonary function change were defined for cotton workers who smoke as 80 ng/m^3 [20], but the results from a larger study of a mixed population (with reference to smoking history and prior cotton mill work) indicated a calculated threshold of as little as 9 ng/m^3 [21]. The results of Castellan et al. [21] suggest that measuring endotoxin concentration in air is a much more reliable means of assessing the risk of an acute airway response to exposure to cotton dust than measuring the mass concentration of the airborne dust.

Industrial and non-industrial environments that require humidification of the air can provide sources of exposure to gram-negative bacteria and endotoxins [23]. One such case study of humidifier fever in office building workers, a traditionally non-dusty environment, demonstrated the presence of gram-negative bacteria in the air [11]. Although endotoxin levels were not quantified, the investigators implied that endotoxins were responsible for the illness. In a more traditional industrial situation, that of a textile facility, cases of lung disease similar to hypersensitivity pneumonitis or humidifier fever were reported [24, 25]. Extensive investigations of the facility demonstrated the putative agent of the lung disease as endotoxin from a species of _Cytophaga_, a microbial inhabitant of soil and fresh water.

Exposures to gram-negative bacteria and endotoxins can occur in unlikely environments as well. During a college fraternity party, where 4-6 inches of straw were spread over the floor in a poorly ventilated room, attendees became ill with clinical and epidemiologic characteristics of organic dust toxic syndrome [26] an acute self-limiting febrile reaction after inhaling high concentrations of agricultural dusts. The investigators did not sample for endotoxins or bacteria. However, our experience with dried silage suggests a high possibility of exposures to large amounts of endotoxins was high in this situation [27].

SAMPLING, COLLECTION, AND TRANSPORT, AND ANALYSIS

Bulk materials, water, and airborne dusts can be tested for the presence of endotoxins. Respirable, total, vertically elutriated or multi-staged size fractionated dusts can be used also for analyses. Sampling times and flow rates will depend upon the environment to be studied. Polyvinyl chloride (PVC) filters, PVC-co-polymer filters, and glass fiber filters are used commonly, although the type of filter that is used to collect airborne dusts appears to be of little consequence [28]. Standard safeguards in handling the filters before and after dust collection should be followed. The container that the bulk sample or dust is in should not be a source of extraneous endotoxins. For this reason, and because dusts may escape from the filters during transport, it is recommended that, after post-weighing, the filters be placed in 50 ml sterile plastic conical centrifuge tubes with screw caps. They then may be mailed or transported. Longer transport times or transport under unique situations may require additional handling such as freezing or cold shipment to reduce the possibility of significant growth of gram-negative bacteria. In the laboratory, the extracts can be made directly in the transport tube, and no dusts will be lost.

In general, investigators rely on the Limulus amebocyte lysate gelation test method for quantitation of endotoxins [29]. However, modifications of the Limulus assay are used, and different techniques are being tried. Haglind and Rylander [20] used the Limulus tube clot technique in their study of the controlled exposure of human subjects to cotton dust, while Castellan et al. [21] used a spectrophotometric modification of the Limulus gel technique in their study. Analyses of environmental dusts for endotoxin contamination were performed with other techniques such as: a microtiter modification of the Limulus test [30]; kinetic Limulus test [31, 32]; gas chromatography [33]; and the quantitative chromogenic modification of the Limulus test [34, 35].

In an attempt to develop scientific standardization of the test used to determine endotoxin content in environmental dusts, the Workgroup on Agents in Organic Dusts in the Farm Environment recommended the use of the quantitative chromogenic modification of the Limulus amebocyte lysate test [28]. Because of its accuracy and reproducibility, the same technique was recommended at an international symposium on work-related respiratory disorders among farmers [36].

When working with the chromogenic modification of the Limulus assay, recent investigations into sample handling and treatment found that sterile, non-pyrogenic water is the extraction medium of choice [37]. In our laboratories, the usual extraction time is 60 min at room temperature, followed by centrifugation of the decanted fluid for 10 min at 1000 x G.

Only a few controlled studies approach the difficult task of relating endotoxin exposure with disease or altered reactivities. Major studies of controlled exposures to endotoxins were done as part of research related to cotton dust exposure [19, 20, 21]. The dust exposures were similar to the workplace situations, but the exposures were not to endotoxin alone. Dust component interactions that could not be tested may have been active in these studies. Nonetheless, two thresholds of endotoxin exposure for zero acute pulmonary function change were 80 ng/m^3 [20] and 9 ng/m^3 [21]. Whether these numbers can be extrapolated to other dust exposures and whether these figures for acute pulmonary function change can provide insight into chronic lung impairment are yet undefined. However, an actual cotton mill study was performed in The People's Republic of China [22], and an association between the endotoxin levels in the mills and chronic lung impairment was found. Additionally, whether individual susceptibility to the effects of endotoxins is found in exposed workers is likewise undefined.

INTERVENTION

In the model cardroom, effective water washing of cotton before the carding process reduces the endotoxin content in the lint and in the airborne dusts that are generated during carding [38, 39]. Although variations in washing treatments may affect the efficacy of endotoxin removal [38, 39], human exposure studies demonstrated that the acute bronchoconstrictor potency of the card-generated dusts can be reduced by mild washing of cottons [40].

In other areas of agriculture, recommendations have been made that farm workers should use National Institute for Occupational Safety and Health/Mine Safety and Health Administration (NIOSH/MSHA) approved dust and mist respiratory protective equipment while unloading silos [41]. Controlled studies of the protective effects of such equipment when used in endotoxin-contaminated air should be conducted and evaluated before a more general comment can be made.

Recommendations for the control of microbiological hazards associated with air conditioning and ventilation systems have been published [42, 43], and full discussion of this subject is beyond the scope of this paper.

SPECIAL TOPICS

Results from endotoxin-related research suggest that gram-negative bacterial endotoxins may differ in their relative toxicities. Lipopolysacccharides from different bacteria which were isolated from cottons differed in chemical composition and in their pulmonary toxicities for laboratory animals [44, 45]. Further animal inhalation studies that used markers of inflammation as a measure of relative toxicity of inhaled gram-negative bacteria concluded that the tested bacteria differed in their toxicity [46]. It is probable, therefore, that different dusts or endotoxin-contaminated air may differ in their relative toxicities based on the species of contaminating organisms.

Many questions concerning endotoxins remain unanswered in this relatively recent area of research. In addition to the relative toxicity question, questions remain regarding the role of endotoxins as adjuvants to the toxicity of other agents. Can they act biologically in an additive or synergistic manner? Do interactions with other microorganisms amplify the growth of gram-negative bacteria and the release of endotoxins into the environment? Do endotoxins from genetically engineered bacteria differ in their humoral and cellular interactions? While these questions are not meant to be all-inclusive, they do illustrate the development of different scientific questions relating to endotoxins and their release into the environment.

CONCLUSION

Endotoxins from gram-negative bacteria are potent biologic agents that can be found throughout the world in soil, water, and dusts. They are associated with acute pulmonary function decrements when inhaled along with cotton dusts, and their presence in other agricultural and non-agricultural dusts is a major research area for the scientific community. Even environments that are considered traditionally to be non-dusty can be a source of airborne endotoxins with potentially deleterious effects on the health of exposed individuals.

ACKNOWLEDGEMENTS

The author thanks Ms. Beverly Carter for her help in preparing this manuscript. Mention of company names or products does not constitute endorsement by the National Institute for Occupational Safety and Health.

REFERENCES

[1] Windholz, M., Budvari, S., Stroumtsos, L. Y. and Festig, M. N., Ed., The Merck Index. Ninth Edition, Merck and Company, Rahway, NJ, 1976, p. 469.

[2] Morrison, D. C. and Ulevitch, R. J., "The Effects of Bacterial Endotoxins on Host Mediation Systems," American Journal of Pathology, Vol. 93, 1978, pp. 527-617.

[3] Galanos, C., Freudenberg, M. A., Luderitz O., Rietschel, E. T. and Westphal, O., "Chemical, Physicochemical and Biological Properties of Bacterial Lipopolysaccharides," in Biomedical Applications of the Horseshoe Crab (Limulidae), Alan R. Liss, Inc., New York, 1979, pp. 321-332.

[4] Rietschel, E. T., Brade, H., Brade, L., Kaca, W., Kawahara, K., Lindner, B., Luderitz, T., Tomita, T., Schade, U., Seydel, U. and Zaringer, U., "Newer Aspects of the Chemical Structure and Biological Activity of Bacterial Endotoxins," in Bacterial Endotoxins: Structure, Biomedical Significance, and Detection with the Limulus Amebocyte Lysate Test, Alan R. Liss, Inc., New York, 1985, pp. 31-50.

[5] Joklik, W. K., Willett, H. P., Amos, D. B. and Wilfert, C. M., Zinsser Microbiology. Nineteenth Edition, Appleton and Lange, Norwalk, CT, 1988, pp. 70-72.

[6] Bradley, S. G., "Cellular and Molecular Mechanisms of Action of Bacterial Endotoxins," Annual Review of Microbiology, Vol. 33, 1979, pp. 67-94.

[7] Duncan, R. L., Hoffman, J., Tesh, V. L. and Morrison, D. C., "Immunologic Activity of Lipopolysaccharides Released from Macrophages after the Uptake of Intact E. coli in Vitro," Journal of Immunology, Vol. 136, 1986, pp. 2924-2929.
[8] Rylander, R., Bake, B., and Fischer, J., "Reactions in Humans after Exposure to Pure and Cell Bound Endotoxin," in Proceedings of the Tenth Cotton Dust Research Conference, National Cotton Council, Memphis, TN, 1986, pp. 114-115.
[9] Dutkiewicz, J., "Bacteria in Farming Environment," European Journal of Respiratory Diseases, Vol. 71, 1987, pp. 71-88.
[10] Dutkiewicz, J., Jablonski, L. and Olenchock, S. A., "Occupational Biohazards: A Review," American Journal of Industrial Medicine, Vol. 14, 1988, pp. 605-623.
[11] Rylander, R., Haglind, P., Lundholm, M., Mattsby, I. and Stenqvist, K., "Humidifier Fever and Endotoxin Exposure," Clinical Allergy, Vol. 8, 1978, pp. 511-516.
[12] Rylander, R. and Snella, M-C., "Endotoxins and the Lung: Cellular Reactions and Risk for Disease," Progress in Allergy, Vol. 33, 1983, pp. 332-344.
[13] Davis, W. B., Barsoum, I. S., Ramwell, P. W. and Yeager, H. Jr., "Human Alveolar Macrophages: Effects of Endotoxin in Vitro," Infection and Immunity, Vol. 30, 1980, pp. 753-758.
[14] DeLucca, A. J. II, Brogden, K. A. and Engen, R., "Enterobacter Agglomerans Lipopolysaccharide-Induced Changes in Pulmonary Surfactant as a Factor in the Pathogenesis of Byssinosis," Journal of Clinical Microbiology, Vol. 26, 1988, pp. 778-780.
[15] Lundholm, M. and Rylander, R., "Work Related Symptoms Among Sewage Workers," British Journal of Industrial Medicine, Vol. 40, 1983, pp. 325-329.
[16] Matson, S. C., Swanson, M. C., Reed, C. E. and Yunginger, J. W., "IgE and IgG-Immune Mechanisms Do Not Mediate Occupation-Related Respiratory or Systemic Symptoms in Hog Farmers," The Journal of Allergy and Clinical Immunology, Vol. 72, 1983, pp. 299-304.
[17] Donham, K. J., Zavala, D. C. and Merchant, J. A., "Respiratory Symptoms and Lung Function Among Workers in Swine Confinement Buildings: A Cross-Sectional Epidemiological Study," Archives of Environmental Health, Vol. 39, 1984, pp. 96-101.
[18] Thelin, A., Tegler, O. and Rylander, R., "Lung Reactions During Poultry Handling Related to Dust and Bacterial Endotoxin Levels," European Journal of Respiratory Disease, Vol. 65, 1984, pp. 266-271.
[19] Castellan, R. M., Olenchock, S. A., Hankinson, J. L., Millner, P. D., Cocke, J. B., Bragg, C. K., Perkins, H. H. Jr. and Jacobs, R. R., "Acute Bronchoconstriction Induced by Cotton Dust: Dose-Related Responses to Endotoxin and Other Dust Factors," Annals of Internal Medicine, Vol. 101, 1984, pp. 157-163.
[20] Haglind, P. and Rylander, R., "Exposure to Cotton Dust in an Experimental Cardroom," British Journal of Industrial Medicine, Vol. 41, 1984, pp. 340-345.
[21] Castellan, R. M., Olenchock, S. A., Kinsley, K. B. and Hankinson, J. L., "Inhaled Endotoxin and Decreased Spirometric Values," New England Journal of Medicine, Vol. 317, 1987, pp. 605-610.

[22] Kennedy, S. M., Christiani, D. C., Eisen, E. A., Wegman, D. H., Greaves, I. A., Olenchock, S. A., Ye, T-T. and Lu, P-L., "Cotton Dust and Endotoxin Exposure-Response Relationships in Cotton Textile Workers," American Review of Respiratory Disease, Vol. 135, 1987, pp. 194-200.

[23] Nordman, H., "Humidifier Syndrome," in Occupational Lung Disease, Raven Press, New York, 1984, pp. 97-107.

[24] Flaherty, D. K., Deck, F. H., Cooper, J., Bishop, K., Winzenburger, P. A., Smith, L. R., Bynum, L. and Witmer, W. B., "Bacterial Endotoxin Isolated from a Water Spray Air Humidification System as a Putative Agent of Occupation-Related Lung Disease," Infection and Immunity, Vol. 43, 1984, pp. 206-212.

[25] Flaherty, D. K., Deck, F. H., Hood, M. A., Liebert, C., Singleton, F., Winzenburger, P. A., Bishop, K., Smith, L. R., Bynum, L. M. and Witmer, W. B., "A Cytophaga Species Endotoxin as a Putative Agent of Occupation-Related Lung Disease," Infection and Immunity, Vol. 43, 1984, pp. 213-216.

[26] Brinton, W. T., Vastbinder, E. E., Greene, J. W., Marx, J. J. Jr., Hutcheson, R. H. and Schaffner, W., "An Outbreak of Organic Dust Toxic Syndrome in a College Fraternity," Journal of the American Medical Association, Vol. 258, 1987, pp. 1210-1212.

[27] Olenchock, S. A., May, J. J, Pratt, D. S. and Morey, P. R., "Occupational Exposures to Airborne Endotoxins in Agriculture," in Detection of Bacterial Endotoxins with the Limulus Amebocyte Lysate Test, Alan R. Liss, Inc., New York, 1987, pp. 475-487.

[28] Popendorf, W., "Report on Agents," American Journal of Industrial Medicine, Vol. 10, 1986, pp. 251-259.

[29] Rylander, R. and Burrell, R., "Conference Report," Annals of Occupational Hygiene, Vol. 32, 1988, pp. 553-556.

[30] Fischer, J. J., Morey, P. R. and Foarde, K. K., "The Distribution of Gram-Negative Bacteria and Endotoxin on Raw Cotton Components," American Industrial Hygiene Association Journal, Vol. 47, 1986, pp. 421-426.

[31] Millner, P. D., Perkins, H. H. Jr. and Harrison, R. E., "Methods for Assessment of the Endotoxic Respirable Dust Potential of Baled Cotton," in Proceedings of the Twelfth Cotton Dust Research Conference, National Cotton Council, Memphis, TN, 1988, pp. 3-5.

[32] Milton, D. Gere, R., Feldman, H. and Greaves, I., "A Precise, Sensitive Limulus Test for Airborne Endotoxin," American Review of Respiratory Disease, Vol. 139, 1989, p. A387.

[33] Kirschner, D., Que Hee, S. S. and Clark, C. S., "Method for Detecting the 3-Hydroxymyristic Acid Component of the Endotoxins of Gram-Negative Bacteria in Compost Samples," American Industrial Hygiene Association Journal, Vol. 46, 1985, pp. 741-746.

[34] Attwood, P., Versloot, P., Heederik, D., De Wit, R. and Boleij, J. S. M., "Assessment of Dust and Endotoxin Levels in the Working Environment of Dutch Pig Farmers: A Preliminary Study," Annals of Occupational Hygiene, Vol. 30, 1986, pp. 201-208.

[35] Olenchock, S. A., "Quantitation of Airborne Endotoxin Levels in Various Occupational Environments," Scandinavian Journal of Work, Environment, and Health, Vol. 14, 1988, pp. 72-73.

[36] Rylander, R., "Role of Endotoxins in the Pathogenesis of Respiratory Disorders," European Journal of Respiratory Disease, Vol. 71, 1987, pp. 136-144.
[37] Olenchock, S. A., Lewis, D. M., Mull, J. C. and Keenan, N. J., "Analysis of Environmental Dusts by the Quantitative Chromogenic Limulus Amebocyte Lysate Test," in Proceedings of Endotoxin Inhalation Workshop, National Cotton Council, Memphis, TN, 1988, pp. 198-199.
[38] Olenchock, S. A., Mull, J. C. and Jones, W. G., "Endotoxins in Cotton: Washing Effects and Size Distribution," American Journal of Industrial Medicine, Vol. 4, 1983, pp. 515-521.
[39] Olenchock, S. A., Millner, P. D., Fischer, J. J., Perkins, H. H. Jr. and Jacobs, R. R., "Microbiology of the Fiber and Airborne Dust from Washed Cotton," in Washed Cotton: Washing Techniques, Processing Characteristics, and Health Effects, United States Department of Agriculture, New Orleans, LA, 1986, pp. 53-74.
[40] Castellan, R. M., "Evaluation of Acute Human Airway Toxicity of Standard and Washed Cotton Dusts," in Washed Cotton: Washing Techniques, Processing Characteristics, and Health Effects, United States Department of Agriculture, New Orleans, LA, 1986, pp. 41-52.
[41] May, J. J., Pratt, D. S., Stallones, L., Morey, P. R., Olenchock, S. A., Deep, I. W. and Bennett, G. A., "A Study of Dust Generated During Silo Opening and Its Physiologic Effects on Workers," in Health and Safety in Agriculture, CRC Press, Inc., Boca Raton, FL, 1989, pp. 76-79.
[42] Ager, B. P. and Tickner, J. A., "The Control of Microbiological Hazards Associated with Air-Conditioning and Ventilation Systems," Annals of Occupational Hygiene, Vol. 27, 1983, pp. 341-358.
[43] Morey, P. R., Jones, W. G., Clere, J. L. and Sorenson, W. G., "Studies on Sources of Airborne Microorganisms and on Indoor Air Quality in a Large Office Building," in IAQ '86: Managing Indoor Air for Health and Energy Conservation, American Society for Heating, Refrigerating and Air-Conditioning Engineers, Inc., Atlanta, GA, 1986, pp. 500-509.
[44] Helander, I., Salkinoja-Salonen, M. and Rylander, R., "Chemical Structure and Inhalation Toxicity of Lipopolysaccharides from Bacteria on Cotton," Infection and Immunity, Vol. 29, 1980, pp. 859-862.
[45] Helander, I., Saxen, H., Salkinoja-Salonen, M. and Rylander, R., "Pulmonary Toxicity of Endotoxins: Comparison of Lipopolysaccharides from Various Bacterial Species," Infection and Immunity, Vol. 35, 1982, pp. 528-532.
[46] Baseler, M. W., Fogelmark, B. and Burrell, R., "Differential Toxicity of Inhaled Gram-Negative Bacteria," Infection and Immunity, Vol. 40, 1983, pp. 133-138.

DISCUSSION

Could you address the limitations of the LAL assay in detecting endotoxin from different gram-negative bacteria?

Would some LPS not be detected by LAL, yet still have biological activity on humans?

CLOSURE

Endotoxins from various gram-negative bacteria will be detected by the *Limulus* amebocyte lysate (LAL) assay. The test may not differentiate necessarily endotoxins with varying degrees of toxicities when assayed for different biological properties. It is known, for example, that various gram-negative bacteria are not equally toxic when inhaled by animals.

Questions have been made recently concerning endotoxins which remain associated to the bacterial cells and whether their presence will be estimated accurately by the LAL assay. Research relating results from dose and response measurements should lead to resolution of some of these types of questions.

DISCUSSION

What is the recommended method for air sampling of endotoxin?

The outline mentions a hypothetical case of endotoxin-related humidifier fever. What would be the best way to sample and analyze airborne and bulk water samples?

CLOSURE

Standard methods for air sampling for endotoxins have not been defined. In many ways the methods would depend on the environment to be measured and the type and quantity of airborne materials. The manuscript addresses certain methods which have proven to be successful, and various groups have endorsed the chromogenic modification of the *Limulus* amebocyte lysate test for quantitating endotoxins in environmental samples.

DISCUSSION

Do all bacteria contain endotoxin? What determines their presence and release?

CLOSURE

Gram-negative bacteria contain endotoxins. They are released after bacterial lysis and during active cell growth. From this we know, therefore, that endotoxins can be found in environments in which viable gram-negative bacteria cannot be demonstrated.

DISCUSSION

Do you see any role of protozoa in ingestion of gram-negative bacteria and enrichment for endotoxins?

CLOSURE

It would seem that any situation in which the growth of gram-negative bacteria is increased could result in the formation and presence of endotoxins.

DISCUSSION

Are there any data on levels of endotoxin and/or human effects in homes downwind from sewage treatment plants?

CLOSURE

Published data do exist for exposures and health-related effects in workers in waste treatment and sewage sludge operations both in the United States and in Sweden. I am not aware if those investigators examined areas beyond the occupational environments.

DISCUSSION

What level of endotoxins in bulk carpet or HVAC systems would be considered "contaminated" regardless of the presence of adverse health symptoms?

Have you ever measured endotoxin concentrations in cooling tower reservoir water, especially after biocide treatment?

How do you decontaminate?

CLOSURE

Our experience in these areas is limited to only one preliminary study of carpet in which we found 4.6×10^3 colony forming units of gram-negative bacteria per square inch. The endotoxin level was 7.58×10^4 Endotoxin Units per square inch. It is difficult to interpret these data from such a limited study. I am not aware of an agreed upon action level for this type of contamination.

Removal of gram-negative bacteria from contaminated materials would remove the source of new endotoxin contamination. Engineering controls and decontamination procedures are beyond the scope of this paper.

Bruce B. Jarvis

MYCOTOXINS AND INDOOR AIR QUALITY

REFERENCE: Jarvis, B. B., "Mycotoxins and Indoor Air Quality,"Biological Contaminants in Indoor Environments, ASTM STP 1071, Philip R. Morey, James C. Feeley, Sr., James A. Otten, Editors, American Society for Testing and Materials, Philadelphia, 1990.

ABSTRACT: Mycotoxins are secondary metabolites produced by fungi which pose a hazard to the health of humans and animals. Numerous cases of intoxication through ingestion of contaminated food and feed have been reported, but little is known about the airborne threat of toxigenic fungal spores. Although toxigenic isolates of fungi are found indoors, they are relatively uncommon. Literature reports suggest that mycotoxins may play a role in the symptoms experienced by those occupants of buildings which are heavily contaminated by certain fungal species.

KEYWORDS: *Stachybotrys, Aspergillus, Penicillium*, trichothecenes, air sampling, molds, mycotoxicosis, immunosuppressants

INTRODUCTION

Fungi are ubiquitous in nature and serve a central role in the ecology of this planet, for they are the agents principally responsible for the breakdown of organic matter. Fungi in this role are known as saprophytes though some may act as pathogens in both plants and animals.

Systemic fungal infections in humans are relatively rare. In North America, the most common mycoses are probably those caused by *Candida albicans* which causes vaginal, throat and, occasionally, systemic infections and by *Aspergillus fumigatus* (and some other fungi), which causes aspergillosis. In immunocompromised individuals (burn victims, AIDS patients, those on certain medications) organ transplant recipients, etc., a wider variety of mycoses from these and various opportunistic molds are possible. The second problem well known to medicine involving the inhalation of fungal spores is allergy. The spores or mycelial fragments of fungi contain allergens. Provided the species causing the allergy is one of the 40 or so for which allergy test kits can be purchased from medical supply companies, this can be diagnosed with little difficulty. Unfortunately, there are >100,000 species of molds, at least 1,000 of which could be described as common in North America, for which allergy tests are not available. The third problem recognized by medical doctors caused by the inhalation of fungi is so-called hypersensitivity pneumonitis. This involves the reaction to the inhalation of massive amounts of organic material including spores [1,2].

About 15 years ago, doctors working in rural areas where grain is handled reported a syndrome called "atypical farmer's lung." Wherein, what should have come and gone as an occurrence of hypersensitivity pneumonitis, proved to be much more serious than usual [3]. Several research groups now associate this phenomenon with the inhalation of mycotoxins in dusts or spores [4]. The spores of toxigenic fungi contain large concentrations of whatever toxins the fungus can produce. *A. flavus* conidia, for example, can contain >600 ppm aflatoxins [5].

Mycotoxins and mycotoxicoses nearly always arise in the context of agriculture, usually through a contamination of cereals and grains brought about either through fungal pathogenic activity in the fields or from postharvest saprophytic fungal activity. Studies of this area increased dramatically in the West following the discovery of the *Aspergillus*-produced carcinogenic aflatoxins in England in 1960. The Japanese and Eastern Europeans were well-aware of the serious hazards associated with mycotoxins many years before [6].

The mycotoxicologist has a difficult time sorting out just what is responsible for a given toxic episode since often the environmental sample (e.g. contaminated grain) has been disposed. When the suspect samples are available, they usually are insufficient for either chemical isolation or toxicological work and usually serve only as a source of the suspect toxigenic fungus. Even then, interpretation problems occur since not only is it common for different genera of fungi to be isolated from the same sample, but also several species of the same fungal genus are often isolated. In addition, different isolates of the same species may behave quite differently with respect to mycotoxin production, some producing toxins, others not. It is difficult to reproduce the correct culture conditions *in vitro* to mimic what has occurred naturally. Precise knowledge of growth conditions optimal for the production of each toxin is required. This and the relevant chemical standards are usually lacking. Additionally, laboratory cultures are of a single organism, while the environment always will provide mixed cultures.

There are two additional problems facing the investigating mycotoxicologist: (1) there is not only a large number of different mycotoxins, but they also represent a huge number of structure types ranging from oligiopeptides (molecular weight > 2000) to such simple molecules as moniliformin ($C_4H_2O_3$), and (2) they differ in their biological properties. This means that the investigator is forced to make "an educated guess" as to which mycotoxin(s) might be present and assay accordingly. He will be guided, in part, by both the symptoms of the afflicted and microbiological survey of the environmental material.

The most common fungi involved in mycotoxicoses are *Fusarium, Penicillium* and *Aspergillus*. In the case of the two latter genera, only certain species such as *P. viridicatum, P. cyclopium, A. flavus,* and *A. parasiticus* appear to be of general importance. These *Aspergillus* species produce a series of potent liver carcinogens, the aflatoxins, which are of concern, as airborne toxins, for those who handle material contaminated by these fungi. Inhalation exposure to aflatoxin-containing spores has been demonstrated to result in elevated liver cancer risk for those who handle material contaminated by these fungi [5,7] and safety measures for those dealing with such material have been published [8]. Although these species of *Aspergillus* and a number of *Fusarium* species are commonly associated with mycotoxicoses involving agricultural products, they are rarely found in offices and homes. However, toxigenic species of *Penicillium* are commonly found in indoor air.

Airborne Mycotoxins and Sick Buildings

Fungal spores are to be found in virtually every home and building unless elaborate procedures (e.g. HEPA filtration system) are in place to make it otherwise. The level of viable spores/m^3 in normal homes and buildings is a function of the outdoor airspora. In climates with snow cover, this can approach 10 in winter and be on the order of thousands in summer. Buildings which are associated with people suffering obvious (and often not-so-obvious) mold problems have spore counts from an unknown lower limit to hundreds of thousands of spores/m^3. In particular, constant exposure to spores of toxigenic molds, especially those with potent toxins might only need to be modest for health effects to occur. Studies of the inhalation exposure to certain pure mycotoxins (trichothecenes) have revealed that inhalation exposure can induce a toxicosis at $^1/_{10}$ [9] to less than $^1/_{20}$ [10] of the intervenous dose, though these data vary with the specific animal species under study [11]. This does not take into account the synergy of various toxins that could occur in terms of inhalation spores of several toxigenic species.

It is important to note that most methods of variable air sampling (e.g. Cascade impactors, slit to agar sampler, sieve impactors, centrifugal impactors) result in spores impacting on agar plates. The spore count is then arrived at by counting the colonies of the various fungi that appear after a suitable growth period. Although this method gives a qualitative picture of the significant elevation of airborne fungal spores in homes with mold problems, there are a number of serious limitations.

1) Spores need not be viable to elicit a toxic effect, i.e. dead spores can cause allergic reactions just as readily as viables ones, and mycotoxins can be present regardless of whether the spores are viable or not.

2) Since some fungal spores are longer lived than others and the sedimentation rate of spores of different species is quite different, to get a truly accurate picture, a combination of techniques should be employed to obtain counts of both viable and nonviable fungal spores [12,13].

3) Most studies make no attempt at determining the fungal species, i.e. usually only the genus is determined. Not only is there a great deal of variation between species of the same genus with respect to mycotoxin production, but different isolates of the same species may exhibit quite different patterns of mycotoxin production. Furthermore, some fungal species may produce toxins under some growth conditions and not under other conditions. Potentially, strains of toxigenic molds will make some or all of the toxins known from that species.

4) The nutritional requirements of fungi vary dramatically. Those fungi which are fastidious and/or compete poorly against faster growing species, will always be underrepresented on the culture plates.

5) The specific source of mold contamination in a building with respect to the air flow in and through the building can induce severe problems in quantifying molds. For example, in a large building, sampling just after the heating, ventilation and air conditioning (HVAC) system has been shut down for the weekend will give a different answer than sampling early on Monday morning. Sampling a house with windows open will give a different result than sampling with a down-wind, upstairs window open.

The most common fungi found in buildings are *Penicillium, Alternaria*, and *Cladosporium* and to a lesser extent *Acremonium, Aspergillus* (but not the aflatoxin-producing species), *Trichoderma, Drechslera,* and *Epicoccum.* All but the latter two, commonly have several species which are well-known sources of mycotoxins.

Various studies have associated illness by indoor exposure to toxigenic fungi such as *Trichoderma viride, Stachybotrys atra* and *Penicillium viridicatum* but cause-effect has not been proven unequivocally [2,3]. As noted previously, occupational exposure to *A. flavus/A. parasiticus* spores is the only definitive circumstance of mycotoxicosis through inhalation exposure [2,7].

Stachybotrys: A Case Study

Stachybotrys atra (*chartarum*) is responsible for a number of toxic episodes in farm animals (horses, sheep and cattle). There is a large volume of literature on this subject [14], but it is generally unappreciated in the West, because the vast majority of cases of stachybotryotoxicoses have been reported from Russia and Eastern Europe; only a few cases have been reported from other areas [15]. Whereas thousands of livestock (horses are particularly sensitive) have been lost in Russia and Eastern Europe this century, no cases of stachybotryotoxicosis for North American farms have appeared in the literature.

The Ukraine in particular has had a long and unfortunate history of equine stachybotryotoxicosis. There is even a reference by Khrushchev [16] to an outbreak of mysterious deaths in Ukrainian horses in the early 1940's, where Stalin wanted to have the local veterinarians shot, since he felt that they were poisoning the animals. However, mycologists were able to trace the origin of the problem to a fungus, most certainly *S. atra*, growing on the straw being eaten by the horses.

The clinical symptoms and pathology of stachybotryotoxisis varies depending upon the dosage. High levels lead to rapid death, which is accompanied by massive hemorrhaging, both internal and external. The first response that is evident is necrotic lesions around the area of the mouth and general, often severe, conjunctivitis. Within a few days, leukocytosis and eventually leukopenia develop with commensurate lowering of the total leukocyte count. Blood clotting is inhibited and in some cases, the blood may fail to coagulate. The immune system clearly becomes compromised since afflicted animals often suffer septicemia and fall prey to a number of opportunistic infections [15].

Stachybotryotoxicosis in man is far less common and no lethal cases have been reported. The cases almost always arise through handling the contaminated straw and hay that is poisoning farm animals [17,18]. The most consistent symptom is dermatitis (characteristic of the trichothecene mycotoxins, especially the macrocyclic members *vide infra*). The area around the nose and eyes is particularly sensitive as is also the scrotal area. In addition to a cough, rhinitis and nosebleds, it is particularly relevant to note that an irritated throat, moderate fever, headache, general feebleness and fatigue are general symptoms of stachybotryotoxicosis in humans. Some of these symptoms are common complaints of occupants of "sick" buildings; they also occur in organic dust toxic syndrome [19] and chronic fatigue syndrome [20].

S. atra is uncommon in "healthy" homes and buildings. One study found *Stachybotrys* (species not determined) in 2.9% of 68 homes in southern California, [21] though several of these homes were later shown to have some mold problems. For 51 homes with mold problems with no central electrostatic filtration (CEF) system, the Andersen sampler found 6 homes and the rotorod sampler found ten homes with *Stachybotrys*. In one home equipped with a continuous CEF, the spore count for *Stachybotrys* was 635/m^3, which was twice the mean for any of the other fungal spores found in 7 homes similarly equipped with a continuous CEF.

In a related rotorod study comparing 14 nonproblem control homes with 47 homes with endogenous mold problems, the incidence of *Stachybotrys* in the homes increased from 7% (controls) to 19% in the mold problem homes [21]. The overall level of airborne spores increased by a factor of 3 (1150/m^3 to 3641/m^3); whereas, the levels for *Stachybotrys* spores increase by a factor of 15 (19 to 309). However, since only one of the control homes (1/14) was found to have *Stachybotrys* spores, these data must be interpreted with caution.

A case study has been reported for a home in the Chicago area which had suffered water damage (leaking roof) off and on over a period of several years [22]. During this time, the occupants became increasingly ill, with a variety of complaints: headaches, sore throats, hair loss, flu symptoms, diarrhea, fatigue, dermatitis, and generalized malaise. An air sample (6000 liters) was collected by electrostatic precipitation and shown to contain numerous spores of *S. atra* with few other genera present to any significant extent. This in itself is remarkable since *S. atra*, even in moldy homes, seems to be relatively uncommon. However, note that in the typical studies of airborne spores, only the viable spores that grow out on the media are measured; whereas the above determination was made by microscope examination. Unfortunately, no accurate measurements of airborne spore counts were made in this study.

An important point about *Stachybotrys* needs to be made. This organism is somewhat fastidious and competes very poorly with other fungi on typical agar media. Unless one goes to the trouble to use a cellulose-based agar or moist filter paper medium, where *Stachybotrys* grows reasonably well and most other fungi do not, *Stachybotrys* often will be missed.

The principal areas of contamination in this house were located in a cold air duct and on a ceiling (Celotex®) on the second floor, where moisture from a leaking roof had seeped through. Extracts of material and debris from these areas proved highly toxic to weanling rats and adult mice. From extracts of the ceiling board was isolated a series of highly toxic trichothecene mycotoxins [22]. Since other studies have established that spores of *S. atra* accumulate relatively high concentrations of these toxins [23], it seems reasonable to suggest that many of the medical problems suffered by the occupants of this home were due to stachybotryotoxicosis.

The house was decontaminated by removal and replacement of obviously mold-contaminated ceiling tiles and air ducts. Workman carrying out these tasks, particularly ones dealing with removal of air ducts, rapidly developed signs of trichothecene toxicosis: irrated eyes and throats, and severe skin rashes around the eyes and nose. Workers were obliged to wear protective clothing and respirators during their work.

S. atra is not commonly responsible for health problems arising from exposure to moldy buildings. A survey of 51 Canadian homes [24], 70% of whom had mold problems, detected *S. atra* in only one, in spite of the use of a moist sterile filter paper medium which should have picked up this organism. However, extracts of three of 70 dust and lint samples collected in these homes proved to be highly cytotoxic when bioassayed in HeLa cell cultures. These three samples were toxic at a level of *ca.* 3 µg/ml and came from debris with relatively high *Penicillium viridicatum* and/or

Penicillium spp. counts [24]. This suggests the possibility that mycotoxins may be playing some role in the etiology of a few of these cases.

In a similar study involving 62 dwellings in England and Scotland, *S. atra* was found in 12.8% of the homes and 4.5% of the air samples obtained from 47 dwellings in central Scotland [25]. In those few dwellings where *S. atra* was found, it could occur in as much as 86% of the air samples and make up 38% of all the samples, with a spore count as high as 17,900 spores/m^3.

Sampling and Analysis Methods

The quantitative analysis of airborne spores is a difficult matter for the reasons cited earlier. It would involve a great deal of time and expense to quantitate both viable and non-viable spores, and then, to be statistically significant, sampling would have to be made on several occasions. Even then, there is a potential for missing what might take place in isolated instances: a sudden release of spores which rapidly dissipate. For example, the levels of airborne spores are dramatically elevated by activities such as cleaning (e.g. vacuuming carpets) [24,25] and agitation of heat exchanger surfaces [26]. If people are around at this burst of spore production, they will be exposed, but the investigator, coming later will not detect an increase level of spores.

Air sampling is important to give an overall picture of the general level of fungal contamination, but unless there is a dramatic increase in the level of one of the less commonly encountered fungi (e.g. *Stachybotrys*), it will be difficult to use the data as a guide for further experimentation to address the question of the involvement of mycotoxins. A number of air sampling techniques are available and discussed in detail elsewhere in this volume [27]. In way of a caveat about these various sampling techniques, I point out the general need for proper control experiments, with the apparatuses and techniques, to establish that they indeed do give an accurate picture of not only the CFU's but also an accurate accounting of all fungal particles, viable or not.

Perhaps a better way to proceed would be to vacuum the room and plate out the dust and debris on a series of plates, employing various media which would tend to select out a broad spectrum of fungal types (e.g. cellulose agar at pH 8 or moist filter paper for *Stachybotrys* and *Myrothecium*). In addition, the debris should be extracted and the extracts tested for toxicity, perhaps in animals, but because of the expense, a bank of *in vitro* bioassays might be better. Proper controls should be conducted with dust and debris collected from either similar buildings with no mold problems or from areas of the same building where there is no apparent mold problem.

There is little sense carrying out any chemical analyses [28] unless a suspect organism has been identified. Even then, matters may not simplify since fungi such as the Penicillia produce many different classes of mycotoxins. A standard procedure in cases of mycotoxicoses in animals is to grow the suspect fungus in the laboratory on the natural substrate, feed it to animals to establish the connection between the observed toxicity and the fungus, and finally to isolate and test the pure mycotoxin(s) from the laboratory culture.

There are a number of ambiguities in this approach which would increase many fold were the procedure to be adopted for investigating indoor mold-promoted illnesses in humans. In addition to the many problems associated with establishing the etiology of diseases in humans, there is an important difference between mycotoxicoses arising from fungal contaminated foods and mycotoxicoses arising from molds in buildings: in the former, one often finds a preponderance of one fungal species; whereas, in buildings, it is rare to find one species dominating the environment. Thus, one would often be faced with evaluating a large number of cultures obtained from the contaminated building.

General Procedures

The following is perhaps an idealized case and should be understood in the context of real life situations. People in a building would be expressing general, widespread, and long standing complaints that are not relieved to a significant extent when they are outside the building. These complaints can not be traced to allergies. Allergies often are elicited by relatively minor contacts with fungal spores and often disappear soon after people leave the contaminated building. For example, simply the smell of moldy material or handling objects having only traces of mold will precipitate a very quick response from the sensitized sufferer. It is very unlikely that such responses could be due to mycotoxins since the amount of fungal material would be far too low to produce a measurable toxic response. Also, physicians would be unable to account for the symptoms in terms of known infectious agents. If mycotoxins are involved in acute poisoning, it would be very likely that most of the people who have similar exposures will have similar symptoms. This is not likely to be the case with allergies. On the other hand, if the exposure to mycotoxins is long term and at a subclinical level, matters are much less predictable. Very little is known about such effects in man though animal studies suggest that one consequence of low level chronic exposure may be a general depression of the immune system.

Air samples should be collected, on several different occasions, in those areas where complaints are highest. Control samples should be collected at the same time in those areas which appear to be free of complaints. Although the air samples could be collected either by filtration through a 0.2 μ membrane or by electrostatic precipitation, the amount of material collected would probably be too small to be practical. *In vitro* toxicity tests (e.g. cytotoxicity) could be run on extracts (e.g. alcohol extracts) of the material collected, but it is unlikely that enough spores could be collected in a reasonable time to allow either *in vivo* tests in animals or (even less likely) isolation and chemical characterization of the components of the extracts. Perhaps the most important value in collecting spores in this manner (membrane filter, electrostatic precipitation or Burkard spore trap) is a check to see if the CFU's that appear in a viable impactor (this unit should also be employed, *vide infra*) are an accurate reflection of what is really in the air (see earlier discussion of the limitations in air sampling methods). The Andersen sampler is one of several types available, but, in practice, several types of media should be employed. Ideally, a mycologist should be consulted before and after sampling to give some guidance as to types of media (e.g. cellulose agar at pH 8 is recommended for selecting out *Stachybotrys*) to be used and as to the identification of the genus and (ideally) and species of the fungi isolated.

If mycotoxins are involved, it is likely that there is a substantial growth of the toxigenic organism somewhere in the building, and the most suspect place would be the HVAC system. This should be investigated, paying particular attention to accumulations of debris heavily colonized by fungi. This debris should be collected, extracted (methanol or denatured 95% ethanol typically are suitable), and assayed for toxicity [28]. Again, bioassays must be conducted by those with experience in this area. Also note that personnel collecting these samples should wear respirators and protective clothing. If the samples are moist, they should be air-dried immediately. Samples of obviously contaminated building material also should be collected and analyzed. However, keep in mind that at some point, the relevance of fungi growing on these solid substrates must somehow be ascertained with respect to the possibility that occupants are breathing these fungal materials, i.e., is there evidence that the fungi found on building material or growing in the HVAC system is actually transported into the air?

Once the samples are collected, analyzed for fungal species, and evaluated for toxicity, a decision about chemical analysis needs to be made. Again, expert advice will be very valuable. If the microbial and toxicological analyses point to a specific toxigenic fungus whose toxins are known and available, as standards, the environmental samples can be analyzed by a variety of chromatographic, spectrometric, and immunoassay techniques [28].

The problem becomes considerably more complex if the microbial analysis does not point to a specific organism and yet extracts of the environmental samples prove quite toxic. One would need to obtain in pure culture as many of major fungi found in the building, grow each in culture (rice, corn, or shredded wheat are common solid substrates, and there are a variety of liquid culture media), and evaluate each culture for toxicity. Ideally, one would grow each fungal isolate on a variety of media in order to give as great a chance as possible for toxin production. Note, however, that production of toxins in laboratory cultures can not be taken as proof that this same fungal isolate has produced these toxins in the building. There is no substitute for actually detecting the toxins in the contaminated buildings themselves. But, if toxins are produced in laboratory culture, at least it gives the chemist a specific target for which to assay by the methods available [28].

Conclusions and Future Needs

There is little question that molds play a role in the etiology of some "sick buildings," but at this time, this is just about all we can say with certainty. The problem is perhaps more widespread and complex than is generally appreciated since some illnesses (e.g. chronic fatigue syndrome, Pontiac fever, legionaires' disease) appear to have an as-yet uncharacterized component(s) which suppresses the immune system. A search for a viral origin of this component(s) in chronic fatigue syndrome has failed to turn up any convincing positive evidence [29]. Relevant to this is the fact that many mycotoxins (e.g. trichothecenes) are immunosuppressants [30-33]. We recently have found that *S. atra* produces a series of potent anti-complement compounds in addition to the highly toxic trichothecenes [34]. What role the combination of these potent biogens play in the health of those exposed to *S. atra* (or any other mold) is unknown.

Perhaps the most urgent need is that of educating not only the general public as to the potential hazards of the overgrowth of molds in their homes and offices but also the education of those who investigate such buildings. A more careful and controlled evaluation of mold-infested buildings needs to be conducted. There needs to be determined just what relationships exist between the symptoms of the occupants of these buildings and the specific fungi found therein. This task is made very difficult by the fact that it is often impossible to distinguish between symptoms arising from allergic responses and those caused by mycotoxin intoxication. In addition, it is almost certain that many cases are actually a complex combination of not only allergic responses to several different organisms but quite possibly several different mycotoxins as well.

The two factors in preventing the overgrowth of fungi in buildings are (1) the air exchange rate, and (2) the internal moisture strength (IMS) [24]. Both of these usually can be addressed by routine maintenance [26]. In those buildings suffering heavy mold infestations, surface sterilization (e.g. with hypochlorite) may be insufficient to deal with the long term problem [26], and a thorough cleaning of the heat exchange system may be necessary.

It is clear that the health problems associated with moldy buildings are due to a complex combination of many uncharacterized (perhaps uncharacterizable?) factors. The present data suggest that mycotoxins play an important role in some specific cases, but, in general, their effects are minimal.

ACKNOWLEDGEMENTS

The author acknowledges the assistance of Dr. J. David Miller in the preparation of this manuscript. Support from the National Institutes of Health (NIH-CA-25967) also is gratefully acknowledged.

REFERENCES

[1] Samson, R. A., "Occurrence of Moulds in Modern Living and Work Environments," European Journal Epidemiology, Vol. 1, 1985, pp. 54-61.

[2] Tobin, R. S., Baranowski, E., Gilmon, A. P., Kuiper-Goodman, T., and Miller, J. D., "Significance of Fungi in Indoor Air," Canadian Journal Public Health, Vol. 78, 1987, pp. S1-S32.

[3] Emanuel, J. A., Wenzel, F. J., and Lawton, B. R., "Pulmonary Mycotoxicosis," Chest, Vol. 67, 1975, pp. 293-297.

[4] Flannigan, B., "Mycotoxins in the Air," International Biodeterioration, Vol. 23, 1987, pp. 73-78.

[5] Wicklow, D. T. and Shotwell, O. L., "Intrafungal Distribution of Aflatoxins Among Conidia and Sclerotia of *Aspergillus flavus* and *Aspergillus parasiticus*", Canadian Journal Microbiology, Vol. 29, 1983, pp. 1-5.

[6] Jarvis, B. B., "Mycotoxins - An Overview," in Natural Toxins, Ownby, C. A. and Odell, G. V., eds., Pergamon Press, New York, 1988, pp. 17-29.

[7] Olsen, J. H., Dragsted, L. and Autrup, H., "Cancer Risk and Occupational Exposure to Aflatoxins in Denmark," British Journal Cancer, Vol. 58, 1988, pp. 392-396.

[8] Aflatoxins, Safety Data Sheet No. 17a, DHEW Guidelines for the Laboratory Use of Chemical Substances Posing a Potential Occupational Carcinogenic Risk, November 1979.

[9] Creasia, D. A., Thurmon, J. D., Jones, L. J., Nealley, M. L., York, C. G., Wannemacher, R. W., and Bunner, D. L., "Acute Inhalation Toxicity of T-2 Mycotoxin in Mice," Fundamental and Applied Toxicology, Vol. 8, 1987, pp. 230-235.

[10] Creasia, D. A. and Lambert, R. J., "Acute Respiratory Tract Toxicity of the Trichothecene Mycotoxin, T-2 Toxin," in Trichothecene Mycotoxicosis: Pathophysiologic Effects, Vol. I, Beasley, V. R., ed., CRC Press, Boco Rouge, FL, 1989, pp. 161-170.

[11] Pang, V. F., Lambert, R. J., Felsburg, P. J., Beasley, V. R., Buck, W. B., and Haschek, W. M., "Experimental T-2 Toxicosis in Swine Following Inhalation Exposure: Clinical Signs and Effects on Hematology, Serum Biochemistry, and Immune Response," Fundamental and Applied Toxicology, Vol. 11, 1988, pp. 100-109.

[12] Burge, H. P., Boise, J. R., Rutherford, J. A., and Solomon, W. R., "Comparative Recoveries of Airborne Fungus Spores by Viable and Non-viable Modes of Volumetric Collection," *Mycopathologia*, Vol. 61, 1977, pp. 27-33.

[13] Kozak, P. P., Gallup, J., Cummins, L. H., and Gillman, S. A., "Currently Available Methods for Home Mold Surveys. II. Examples of Problem Homes Surveyed," *Annals of Allergy*, Vol. 45, 1980, pp. 167-176.

[14] Forgacs, J., "Stachybotryotoxicosis," in Microbial Toxins, Vol. VIII, Kadis, S. Ceigler, A., and Ajl, S. J., eds., Academic Press, New York, 1972, pp. 95-128.

[15] Schneider, D. J., Marasas, W. F. O., Kuys, J. C. D., Kriek, N. P. J., and Van Schalkmyk, G. C., "A Field Outbreak of Suspected Stachybotryotoxicosis in Sheep," Journal of the South African Veterinary Association, Vol. 50, 1979, pp. 73-81.

[16] Khrushchev, N., Khrushchev Remembers, Little, Brown and Co., Boston, 1970, pp. 111-113.

[17] le Bars, J. and le Bars, P., "Study of Airborne Spores of *Stachybotrys atra* on Straw: Inhalation Risk," Bulletin de Society Française de Mycologie Medical, Vol. 14, 1985, pp. 321-324.

[18] Andrássy, K., Horváth, I., Lakos, T., and Töke, Zs., "Mass Incidence of Mycotoxicoses in Hajdn-Behar Country," Mykosen, Vol. 23, 1980, pp. 130-133.

[19] Brinton, W. T., Vastbinder, E. E., Greene, J. W., Marx, J. J., Hutcheson, R. H., and Schaffner, W., "An Outbreak of Organic Dust Toxic Syndrome in a College Fraternity," Journal of the American Medical Association, Vol. 258, 1987, pp. 1210-1212.

[20] Swartz, M. N., "The Chronic Fatigue Syndrome - One Entity or Many?," The New England Journal of Medicine, Vol. 319, 1988, pp. 1726-1728.

[21] Kozak, P. P., Jr., Gallup, J., Cummins, L. H., and Gillman, S. A., "Endogenous Mold Exposure: Environment Risk to Atopic and Nonatopic Patients," in Indoor Air and Human Health, Gammage, R. B. and Kaye, S. V., eds., Lewis Publishers, Inc., Chelsea, MI, 1985, pp. 149-170.

[22] Croft, W. A., Jarvis, B. B., and Yatawara, C. S., "Airborne Outbreak of Trichothecene Toxicosis," Atmospheric Environment, Vol. 20, 1986, pp. 548-552.

[23] Sorenson, W. G., Frazer, D. G., Jarvis, B. B., Simpson, J., and Robinson, V. A., "Trichothecene Mycotoxins in Aerosolized Conidia of *Stachybotrys atra*," Applied and Environmental Microbiology, Vol. 53, 1987, pp. 1370-1375.

[24] Miller, J. D., Laflamme, A., Sobol, Y., Lafontaine, P., and Greenhalgh, R., "Fungi and Fungal Products in Some Canadian Houses," International Biodeterioration, Vol. 24, 1988, pp. 103-120.

[25] Hunter, C. A., Grant, C., Flannigan, B., and Bravery, A. F., "Mould in Buildings: the Airspora of Domestic Dwellings," International Biodeterioration, Vol. 24, 1988, pp. 81-101.

[26] Morey, P. R., "Microorganisms in Buildings and HVAC Systems: A Summary of 21 Environmental Studies," Engineering Solutions to Indoor Air Problems, Proceedings of the ASHRAE Conference, IAQ88, April 11-13, 1988, Atlanta, GA, pp. 10-24.

[27] See H. P. Burge in this volume.

[28] Cole, R. J. (Ed.), Modern Methods in the Analysis and Structural Elucidation of Mycotoxins, Academic Press, New York, 1986.

[29] Straus, S. E., Dale, J. K., Tobi, M., Lawley, T., Preble, O., Blaese, R. M., Hallahan, C., and Henle, W., "Acyclovir Treatment of the Chronic Fatigue Syndrome: Lack of Efficacy in a Placebo-Controlled Trial," The New England Journal of Medicine, Vol. 319, 1988, pp. 1692-1698.

[30] Otokawa, M., "Immunological Disorders," in Trichothecenes: Chemical Biological, and Toxicological Aspects, Development in Food Science, Vol. 4, 1983, pp. 163-170.

[31] Hughes, B. J., Hsieh, G. C., Jarvis, B. B., and Sharma, R. P., "Effects of Macrocyclic Trichothecenes Mycotoxins on the Murine Immune System," Archives of Environmental Contamination and Toxicology, Vol. 18, 1989, pp. 388-395.

[32] Taylor, M. J., Pang, V. F., and Beasley, V. R., "The Immunotoxicity of Trichothecene Mycotoxins," in Trichothecene Mycotoxicosis: Pathophysiologic Effects, Vol. II, Beasley, V. R., ed., CRC Press, Boca Rouge, FL 1989, pp. 1-38.

[33] Richard, J. L. and Thurston, J. R., "Effect of Aflatoxin on Phagocytosis of *Aspergillus fumigatus* Spores by Rabbit Alveolar Macrophages," Applied Microbiology, Vol. 30, 1975, pp. 44-47.

[34] Jarvis, B. B. and Salemme, J., unpublished results.

DISCUSSION

Do mycotoxins contaminate surfaces in homes or buildings (objects handled by occupants)?

CLOSURE

Fungi that grow on solid substrates typically do not release mycotoxins; therefore, it is very unlikely that surfaces in buildings will become contaminated with mycotoxins in the absence of obvious fungal growth.

DISCUSSION

(1) Are there any data to document low level exposures to mycotoxins and to detail long term effects of such? (2) Has diarrhea ever been reported to result from long term exposure to airborne mycotoxins or to spores of fungi known to produce mycotoxins?

CLOSURE

I know of no long term studies conducted on the effects of the inhalation of low levels of toxigenic fungi or their mycotoxic metabolites that are likely to be problems in buildings and homes. Most of these long term exposure studies have been conducted orally though there is concern that some workers may be exposed by a respiratory route - see W. G. Sorenson et al. *J. Tox. Environ. Health* **1984**, *14*, 525-533.

DISCUSSION

In sick building syndrome cases where there is no evidence of increased rate of respiratory tract infections, would you consider that *Stachybotrys* toxins might be involved? i.e. does the immunosuppressive effect occur at doses at (or below) the dose causing acute toxic effects (headache, fatigue, etc.)

CLOSURE

Long term exposure to trichothecenes at subacute levels give immunotoxic effects in animals (see M. J. Taylor, V. F. Pang, and V. R. Beasley in Trichothecene Mycotoxicosis: Pathophysiologic Effects, Vol. II, V. R. Beasley, ed., CRC Press, Boca Raton, FL 1989, pp 1-38), but there are no data available for the specific toxins found in *Stachybotrys*. If there are no secondary bacterial infections and no complaints about burning sensations about the eyes and nose, I would say there is little reason to even suspect the involvement of *Stachybotrys*.

DISCUSSION

What effect does humidity have on *S. atra* population in the indoor environmental range of 10-70% RH? If it likes paper, would you expect problems in libraries and offices in particular? What about low level contamination over large surface areas?

CLOSURE

The relative humidity must be high (>70%) for *S. atra* to flourish. In fact, *S. atra* requires very damp conditions, i.e. the material needs to be soaking wet. Normally, when water damage in office or libraries occurs, the water is taken up and the place dried out. In such cases, *S. atra* will not get started. When *S. atra* does take hold, it usually does so in out of the way places and in fairly localized areas rather than growing over large surface areas.

DISCUSSION

Are there *Penicillium* species which are notorious toxin producers? Which ones?

CLOSURE

Indeed there are. Betina (*Dev. Food Sci.* **1984**, *8*, p. 5) lists 29 such toxins from 18 species of *Penicillium*. However, of these species, *P. viridicatum* is probably the only one likely to be contaminating the air we breath in homes and buildings. Unfortunately, this species produces a variety of mycotoxins, although, to date, none have been identified from an indoor environment.

DISCUSSION

What culture media do you suggest for isolating *Stachybotrys*?

CLOSURE

Cellulose agar at pH 8 (see I. A. El-kady et al. *Mycopathologia* **1981**, *76*, 59).

DISCUSSION

Could mycotoxins be collected using traditional absorbents like Tenax, then analyzed via GC/MS? If not, why not, how volatile are these in general?

CLOSURE

Most mycotoxins are not volatile but are found inside the spores or mycelium of the fungi. Typically, the contaminated material is extracted (ethyl alcohol is a good all around solvent for such purposes) and the extract analyzed for the mycotoxins. GC/MS is often the best technique available, but standards must be available for comparison.

DISCUSSION

How do you decontaminate an area where large quantities of organisms have grown? i.e. *Penicillium*, etc.

CLOSURE

If the expense is not too high, it is often best to simply replace the contaminated items. However, it this is not practical, the fungal growth should be scrapped off, and the area washed well with hypochlorite and detergent solutions. Those carrying out this work should wear respirators and protective clothing.

DISCUSSION

If you suspect that *Stachybotrys*/trichothecenes are contaminating a material such as sheetrock or wallpaper, how much material do you need for isolation of trichothecenes? Do you need kilogram, gram, mg or microgram amounts?

CLOSURE

Since the amount of toxin is related to the fungal mass, this will depend upon how heavily the material is contaminated. In the case of heavily contaminated material, a gram or less of material should be sufficient. If the growth is poor, then much larger amounts may be required.

DISCUSSION

How many laboratories are capable of isolating and identifying mycotoxins that may be important in indoor environments? Are there commercial labs that can do this type of work?

CLOSURE

World wide, there are several dozen laboratories which can do these analyses for *specific* classes of toxins. Except in the case of aflatoxins and to a lesser extent the trichothecenes, there are no commercial laboratories which screen for mycotoxins.

DISCUSSION

Toxins released from many fungi such as *Aspergillus flavus* and *Stachybotrys* are extremely toxic; are there specific laboratory guidelines ensuring their safe analysis?

CLOSURE

There are such guidelines for *Aspergillus flavus* - see reference [8] - but not for *Stachybotrys*. In general, work in a hood and never allow this organism or its chemical products to come into contact with skin.

Harriet A. Burge

THE FUTURE

REFERENCE: Burge, H. A., "The Future," Biological Contaminants In Indoor Environments, ASTM STP 1071, Philip R. Morey, James C. Feeley, Sr., James A. Otten, Editors, American Society for Testing and Materials, Philadelphia, 1990.

ABSTRACT: Bioaerosol related research is necessary in establishing exposure/dose/response relationships for the common diseases (infectious, hypersensitivity, toxic). The possibility of synergistic effects should be explored. Tools for reliable and efficient bioaerosol sampling need to be developed and broadly based prevalence studies conducted. Safe means for control, including prevention of sources, and remedial actions need to be developed. We need to explore non-traditional funding options, and work together to share existing limited resources.

The quest for knowledge related to indoor bioaerosols is essentially a quest for a means for the control of the diseases they cause. Although we have made some progress in collecting the existing knowledge that applies to this quest, very little research has been specifically directed toward indoor biological aerosols, and gaps remain that prevent an accurate assessment of the risk to human health imposed by these aerosols. Research is needed in three general areas:

1) exposure/dose, and dose/response relationships for the common biological agents, including host risk factors for each disease, and synergistic effects between biological agents themselves and between these agents and other indoor environmental pollutants

Dr. Burge is an associate research scientist at the University of Michigan Medical Center, R6621 Kresge I, Box 0529, Ann Arbor, MI 48109-0529.

2) the ecology, distribution, and prevalence of biological agents and resulting aerosols in indoor environments, including research on environmental assessment tools

3) effective and safe means for the control of indoor biological pollution.

1) Exposure/dose, dose/response relationships are virtually unknown for any of the common indoor bioaerosols. For example:

a) For Legionella, on which the CDC has spent millions of dollars, the number of airborne bacteria required to initiate an infection is unknown. This means that it is not possible to establish guidelines on acceptable environmental levels of this dangerous and ubiquitous organism. Methods for evaluating airborne exposure will have to be developed, and applied to extensive epidemiologic studies in order to solve this problem.

b) For essentially all the hypersensitivity agents (with the possible exception of house dust mite antigen [1]) levels of environmental antigen required to sensitize and/or cause symptoms are unknown. In addition, host risk factors for hypersensitivity pneumonitis remain unknown. We therefore cannot assess the relative risk of exposure to any environmental antigen (even assuming standardized methods for evaluating exposure were available).

c) A syndrome known as multiple chemical hypersensitivity is being diagnosed and treated in the absence of any direct evidence that the disease exists, or, if it does, that the treatment modes are effective. Nontraditional sensitivity to bioaerosols is also often claimed as part of the syndrome. Objective, scientific studies of this group of symptoms are needed to allow effective treatment directed at the basic causes, whether or not they are related to environmental exposure.

d) With the exception of endotoxin and several of the highly toxic fungus metabolites, the effects on human health of inhalation of bacterial and/or fungal metabolic products have not been studied. We don't know whether or not fungi growing, for example, in a ventilation system contribute to the symptoms commonly known as "sick building syndrome." We do not know whether this kind of fungal growth could contribute to increased rates of infectious disease (by means of immunosuppressive metabolites) or cancer. Both laboratory and field studies are necessary on the kinds of compounds that can become airborne from fungus and bacterial growth in the indoor environment, and on their health effects.

e) Principles utilized in the TEAM (Total Exposure Assessment Methodology) studies done on volatile organic compounds by the EPA need to be applied to bioaerosols [2].

f) The possibility of synergistic effects between bioaerosol types and between bioaerosols and other indoor (and outdoor) air pollutants needs to be addressed. Endotoxin, for example, can act as an adjuvant and may be a necessary co-factor in sensitization in hypersensitivity pneumonitis. Some air pollutants are known to exacerbate the symptoms of asthma. However, is it possible that these same air pollutants could increase the risk of the development of asthma or Legionnaires' disease (for example)?

2) Virtually no organized research has been reported on the prevalence of biological agents in indoor environments. The existing literature consists of case reports, necessarily slanted toward complaint situations. For example:

a) The prevalence of Legionella pneumophila in domestic environments, and factors controlling its prevalence and dissemination remain unclear. Air sampling for Legionella (and for other airborne pathogens such as Histoplasma) remains difficult or impossible. Both are difficult to culture and viability may be damaged during sampling. However, both should be amenable to immunological analysis of appropriately collected samples [3]. A valid air sampling method (possibly based on immunoassay techniques) should be developed, then applied in broadly based field studies designed to pinpoint both sources and mechanisms and rates of release.

b) The species composition of indoor fungal aerosols has not been studied. For example, over three hundred species of Penicillium have been described. As yet, only two studies in narrow geographical areas indicate species composition of in and outdoor aerosols of these common indoor fungi [4, 5]. By far the most common airborne fungus indoors and out (Cladosporium) is rarely identified to species or even grouped on the basis of spore morphology. Without information of this type, immunological assays cannot be developed, and accurate assessment of the role of these fungi in allergic disease cannot be accurately determined. We need to stress the importance of species identification in all bioaerosol surveys related to microorganisms. Basic research is necessary to provide identification tools that can be broadly used.

c) We have no data on the prevalence of toxigenic bacteria or fungi in indoor environments, nor on the prevalence of airborne toxins. In fact, methods that are reliable and reproducible for sampling and assessment of these toxins have not been described. Until we can measure the toxins themselves rather than the producing organisms, accurate estimates of dose and potential risk cannot be determined.

d) While relative humidity has often been suggested as a controlling factor in bioaerosol concentrations in the indoor environment [6], the means whereby increased relative humidity causes such increase has yet to be determined. In addition, the role of relative humidity on human health (especially its role in predisposition to infectious disease) has not been studied in an organized way.

3) While suggestions for the control of indoor bioaerosols are often made and acted upon, very little research has been directed toward demonstrating effective and safe means for the control of indoor biological pollution. For example:

a) Relative humidity control is often suggested to be of prime importance in the prevention of biological contamination. However, the literature is contradictory on actual maximum allowable levels [7, 8]. Different levels are considered important for controlling infectious disease vs preventing surface contamination [9]. Epidemiologic studies relating levels of relative humidity to airborne microbial levels, surface contamination, rates of infectious disease, and rates of sensitization and symptoms related to hypersensitivity disease are necessary.

b) Contaminated fiberglass duct lining is often blamed for building-related illness. However, it is not clear that, under the same circumstances, metal ductwork would have remained clean.

c) Carpeting on cement floors is considered to be a risk for increases in house dust mite levels as well as fungus contamination [1]. However, it is likely that carpeting on a properly designed cement floor could be perfectly safe. Studies should be done that clarify the factors involved in elevated humidity within carpeting.

d) Duct cleaning is often recommended as a cure for building related illness in both homes and office buildings. However, all the evidence that has led to this recommendation is included in anecdotal case reports of severe contamination [9]. Duct cleaning will certainly continue. We need to develop means to compare symptom rates in buildings and homes before and after duct cleaners have done their job.

These are only a few of the many factors involved in indoor biological air pollution and related health risks. We are at the beginning of an exciting new period in research and environmental investigation. Progress will continue to be slow and funding limited, however, until we in the indoor air community make clear to the policy makers the importance of biological pollution. This conference and its resulting publications are an excellent first step. However, influencing budgetary policy is a slow and tricky process, and one at which scientists seem notoriously poor. At least initially, perhaps we can place our reliance on non-traditional sources for research support, including industries that necessarily have strong self-interests. Fortunately, those self-interests are often aligned with science. The fiberglass manufacturers do not need the liability involved if, in fact, their duct liners do pose a significant human health risk. The chemical companies are very interested in developing biocides that are safe and effective.

Meanwhile, all of us interested in bioaerosols need to work together, share new information, and communicate continuously both in conferences such as this and privately, so that the limited resources that are available are not wasted in reinventing the wheel.

REFERENCES

[1] Platts-Mills, T. A. E. and Chapman, M., "Dust mites: immunology, allergic disease, and environmental control," Journal of Allergy and Clinical Immunology, Vol. 80, 1987, pp. 755-775.

[2] Wallace, L. A., The Total Exposure Assessment Methodology (TEAM) Study: Summary and Analysis: Volume I, United States Environmental Protection Agency, Washington, DC, 1987.

[3] Kaplan, W., "Application of the fluorescent antibody technique to the diagnosis and study of histoplasmosis," in Ajello, L., Chick, E. W., et al., Editors, Histoplasmosis, Proceedings of 2nd National Conference, Springfield, IL, Charles C. Thomas, 1971, pp. 327-340.

[4] Muilenberg, M., Burge, H., Sweet, T., Solomon, W., "Penicillium species in and out of doors in Topeka KS," Journal of Allergy and Clinical Immunology, Vol. 85, No. 1, Pt. 2, 1990, pg. 247.

[5] Fradkin, A., Tobin, R. S., Tarlo, S. M., Tucic-Porretta, M., and Malloch, D., "Species identification of airborne molds and its significance for the detection of indoor pollution," Journal of the Air Pollution Control Association, Vol. 37, No. 1, 1987, pp. 51-53.

[6] Kethley, T. W., Cown, W. B., Fincher, E. L., "The nature and composition of experimental bacterial aerosols," Applied Microbiology, Vol. 5, 1957, pg. 1.

[7] American Society of Heating, Refrigerating and Air-conditioning Engineers, Inc., "Ventilation for acceptable indoor air quality," ASHRAE Standard 1989, 1989.

[8] Morey, P. R., Hodgson, M. J., Sorenson, W. G., Kullman, G. J., Rhodes, W. W., Visvesvara, G. S., "Environmental studies in moldy office buildings: biological agents, sources and preventive measures," Annals American Conference of Governmental Industrial Hygienists, Vol. 10, 1984, pp. 21-35.

[9] Crofts, W., Jarvis, B. B., and Yatawara, C. S., "Airborne outbreak of trichothecene toxicosis," Atmospheric Environment, Vol. 20, No. 3, 1986, pp. 549-552.

James C. Feeley, Sr. and Philip R. Morey

TWO CONSULTANTS' VIEWS OF TOMORROW

REFERENCE: Feeley, J. C., Sr. and Morey, P.R., "Two Consultants' Views of Tomorrow," Biological Contaminants in Indoor Environments, ASTM STP 1071, Philip R. Morey, James C. Feeley, Sr., James A. Otten, Editors, American Society for Testing and Materials, Philadelphia, 1990.

ABSTRACT: A number of factors in the future will collectively increase the importance of bioaerosols in indoor environments. The overall human population in the future will be older and hence more susceptible to microbial agents. As a consequence of rising energy costs, less outdoor air may be used to ventilate buildings and more moisture will accumulate indoors. Bioaerosol problems will thus increase in indoor environments. Research will accelerate on finding better and alternative ways to control microbial amplification in porous manmade insulation, heat rejection systems, humidifiers, and other building components. Research on baseline bioaerosol exposures and health risk will accelerate in attempts to develop regulatory guidelines or standards for microbial agents. The commissioning of buildings in the future will involve bioaerosol considerations to minimize the potential for the accumulation, amplification, and dissemination of microbial agents in buildings. Commissioning recommendations will be reflected in consensus engineering standards and future building codes. Specific recommendations for bioaerosol research as related to indoor air quality are enumerated.

KEYWORDS: bioaerosols, future, indoor air quality, commissioning of buildings, research

In a previous paper in this volume [1] we reviewed the current state of knowledge on bioaerosols in nonindustrial indoor environments from the point of view of the investigator, landlord, and tenant. This paper presents our views on the future direction of bioaerosol studies in indoor environments. A number of factors will collectively

Dr. Feeley (deceased September 15, 1989) was President of Pathogen Control Associates, 2204 Hanfred Lane, Suite 100, Tucker, GA 30084. Dr. Morey is Director of Indoor Air Quality Services, Clayton Environmental Consultants, Inc., 151 S. Warner Road, Suite 235, Wayne, PA 19087.

increase the importance of bioaerosols in indoor air quality evaluations.

Susceptibility factors will increase in importance. An aging population will mean that more occupants in buildings in general and especially in nursing homes and health care facilities will be more susceptible to a variety of microbial agents.

Energy costs in the foreseeable future will probably increase. Rising energy costs will stimulate greater energy conservation efforts. In air-conditioned buildings, greater energy conservation often means less dehumidification (or less removal of latent heat) by heat exchangers in heating, ventilation and air-conditioning (HVAC) systems. An increase in moisture indoors will lead to amplification of microbial agents. Increased moisture and microbial proliferation indoors will be more prevalent in buildings in areas of the world, such as the southeast USA, the Far East, and the tropics where air-conditioning is the dominant part of HVAC system operation.

Higher energy costs in the future may also encourage using less outdoor air in HVAC systems. Reduced amounts of outdoor air in indoor environments may result in elevated concentrations of human shed microbial agents such as viruses.

Poor preventive maintenance is probably the single most important factor that leads to microbiological contamination in buildings. The high cost of new construction will result in more existing older buildings. These aging building along with the low priority often assigned to maintenance activities will result in more reservoirs and amplification sites for microorganisms in HVAC systems and occupied spaces. Therefore, bioaerosol problems indoors will increase. Research needs on bioaerosols and the aging of buildings include:

- Determine the kinds and amounts of microbial agents that accumulate in buildings and their HVAC systems as a building ages.

- Document intervention techniques that are effective in preventing accumulation of microbial agents in buildings. These techniques may include better filtration (for example, more efficient filters and use of nonfabric filters) and emphasis on accessibility of mechanical system components to facilitate maintenance.

Research on bioaerosol exposure indoors will accelerate to determine the causes (presumably viruses) of the increased rates of acute respiratory disease among occupants of mechanically ventilated versus naturally ventilated buildings [2]. Future bioaerosol studies will focus on the increased risk of building sickness that British epidemiological studies [3] associate with air-conditioning and humidification in buildings. Finally, the suggestion that the HVAC system itself is the most important source of contaminants that result in odors and unacceptable indoor air [4] should stimulate research on volatile emissions from microorganisms that may be present in mechanical systems. Research needs on bioaerosol exposure in nonindustrial indoor environments include:

- Define the relationship between the concentration of bioaerosols that cause the common cold and outdoor air ventilation rates.

- Determine if elevated concentrations of certain types of bioaerosols such as those that cause the common cold are associated with building or HVAC system parameters such as the presence of variable air volume system terminals, fan coil units, and energy management systems.

- Determine if the presence of moisture in HVAC systems (for example, humidifiers and stagnant water in drain pans) results in elevated concentrations of microorganisms in the occupied zone.

At present, no clear numerical guidelines exist for interpreting bioaerosol analytical data. Interpretation is correctly based on the following parameters [5]:

- The kinds of microorganisms or microbial agents present indoors and in the outdoor air (rank order assessment).

- Medical or laboratory evidence that an adverse health effect is caused by a kind of microorganism. If evidence indicates that illness is caused by a specific microorganism (for example, *Legionella pneumophila* serogroup 1), then the objective of the sampling and the interpretation of analytical results is precise. This, however, is rarely the case.

- Indoor/outdoor concentration ratios of microorganisms or microbial agents. A high ratio suggests that an indoor amplifier is present.

- Concept of reservoir, amplifier, and disseminator [6,7]. In indoor environments, microbial agents may accumulate in reservoirs and their populations may amplify in niches, such as wet porous insulation. However, for an illness to occur, a sufficient amount of the microbial agent must be transported to the breathing zone of a susceptible occupant.

In the 1990's, public health agencies are likely to carry out comprehensive bioaerosol studies in indoor environments in attempting to develop standards and guidelines for bioaerosol exposure. At a minimum, these studies should result in baseline data on "typical" indoor exposure levels to viable and nonviable fungal spores and viruses that cause the common cold as effected by geographic, seasonal, and building variables. Simultaneous bioaerosol and epidemiological studies in nonindustrial indoor environments will likely result in a health effects database required for possible exposure guidelines.

The use of porous accoustical insulation in portions of air handling units that become wet will be studied in the future. Research is required to determine if the almost certain development of microbial amplification sites in wet porous insulation results in the development of building related illnesses such as hypersensitivity pneumonitis and asthma. Research needs on microbial contamination of porous insulation used in HVAC systems are:

- Determine if porous insulation can be placed on external surfaces of air handling units without major accoustical penalties.

- Determine if microbial amplification in porous insulation can be reduced by upgrading the efficiency of upstream filter decks.

- Find cost effective alternatives for removal of latent heat (moisture) from ventilation air that do not involve condensation of water in the mechanical system. Desiccant dehumidification may be an appropriate technology.

Although the use of biocides in cooling towers and evaporative condensers is well established, the use of microbiocidal chemicals in air handling units, humidifiers, and other potential sites of indoor microbial amplification is potentially dangerous. In the latter cases, the beneficial effects of these chemicals must be balanced by their potentially toxic effects on building occupants. Emphasis on biocide utilization in indoor environments in the future will probably involve studies on the incorporation of these chemicals into the actual fabric (fibers) of manmade insulation, filters, carpet, and wall coverings. Future research needs on biocide use in HVAC systems are:

- Determine by case control studies if the use of microbiocidal chemicals is as effective as maintenance without biocides in preventing microbial amplification.

- Demonstrate that biocides used in HVAC systems are not aerosolized into the occupied zone.

- Develop guidelines (protocols) for use of biocides (including disinfectants) in HVAC systems and occupied spaces.

Replacing cooling towers and evaporative condensers with other types of heat rejection systems in the future is probably impractical and not cost effective. However, the increasing importance of water conservation, especially in major urban centers of the world, will require more emphasis on recycling water in building heat rejection systems.

Research is necessary to document the effectiveness of biocides in controlling bacterial amplification in functioning cooling towers. Despite the attention that Legionnaires' disease outbreaks have received in mass media, there is surprisingly little peer-reviewed research on the subject of control of _Legionella_ amplification in functioning cooling towers. Research needs on control of microbial amplification in cooling towers include:

- Determine the effectiveness of biocides as well as other intervention techniques (for example, ozone and hydrogen peroxide) in controlling the selective amplification of _Legionella_ species and serotypes in functioning cooling towers.

- Develop interpretation guidelines on the possible adverse health effects of Legionella species and serotypes found in cooling towers (and other water systems in buildings).

- Develop guidelines to be used by architects for minimal spatial separation of cooling towers from outdoor air inlets of buildings.

In the industrial workplace, outbreaks of hypersensitivity pneumonitis and Pontiac fever [6,7,8] have been associated with exposure to bioaerosols generated by a variety of machines including air washers, humidifiers, and heat rejection equipment. Biocides are used in the water systems of these machines. Only in rare instances has careful biocide usage combined with an excellent preventive maintenance/engineering program been effective in reducing bioaerosol exposure so that indoor bioaerosol concentrations are equivalent to those in outdoor air [9]. Research is needed on methods to control microbial proliferation and associated bioaerosol exposures in these special environments.

Vigilance in the future will be required to detect and identify new respiratory pathogens. During the past 15 years, major new agents that can cause pneumonia (Legionella in the late 1970's; Chlamydia pneumoniae in the 1980's) have been discovered. New respiratory pathogens, possibly including bacteria from potable water, will be identified. An aging population, the increase in the number of immune compromised patients in health care facilities, together with changing environmental parameters involving the trapping of more moisture in buildings, will result in future problems from heretofore unrecognized pathogens.

Bioaerosols will play a greater role in the commissioning of buildings and the codes that govern building construction. Emphasis will be placed on the recognition and elimination of microbial reservoirs and amplifiers at the design stage of building construction or renovation.

The American Society of Heating, Refrigerating and Air-Conditioning Engineers (ASHRAE) Standard 62-1989 [10] recommends the following specific maintenance procedures to reduce conditions conducive to microbial amplification in both HVAC systems and occupied spaces:

- Maintain relative humidities between 30 and 60%. Prevent relative humidity from rising above 70% in air supply ducts.

- Drain pans of air handling units shall be designed for self drainage.

- Air handling and fan coil units shall be designed for easy access for purposes of preventive maintenance, especially for cooling coils/drain pans.

- Because cold water humidifiers and water spray systems are easily contaminated by microbial agents, special care is needed in the maintenance of these devices.

- Entrainment of moisture drift from cooling towers into HVAC system outdoor air inlets must be avoided.

Consensus recommendations on control of microbial amplification and dissemination in ASHRAE Standard 62-1989 require future refinement and incorporation into building operation and maintenance practice as follows:

- Incorporate microbial guidelines contained in ASHRAE Standard 62-1989 into building codes.

- Incorporate detailed maintenance protocols involving control of microbial contaminants in cooling towers, evaporative condensers, water spray systems, and humidifiers into building codes.

Bioaerosol studies in the future will be required to document that proactive maintenance and commissioning procedures do lead to reduced exposures to microbial agents indoors. Major manufacturers of building and HVAC system components, such as saunas, hot water service systems, heat rejection systems, humidifiers, porous manmade insulation in air handling units, are likely to initiate research studies to lessen the potential for microbial amplification in their proprietary products.

Future advances in the development of the state-of-the-art techniques for sampling and analyzing of microorganisms in indoor environments are likely to originate from a number of sources. Two North American organizations, namely, the American Conference of Governmental Industrial Hygienists (ACGIH) and the American Society for Testing and Materials (ASTM) will probably provide much new information on bioaerosols in the future. The ACGIH Bioaerosols Committee has expanded its "Guidelines For Assessment and Sampling of Saprophytic Bioaerosols in the Indoor Environment" [11] to include additional classes of agents. The proceedings of this ASTM Conference on Biological Contaminants in Indoor Environments will hopefully stimulate new thought on sampling and analytical protocols for a wide variety of microbiological agents, such as bacteria, fungi, protozoa, viruses, and microbial toxins.

REFERENCES

[1] Morey, P. R. and J. C. Feeley, Sr., "The Landlord, Tenant, and Investigator: Their Needs, Concerns, and Viewpoints," Conference on Biological Contaminants in Indoor Environments, ASTM, STP 1071, P. R. Morey, J. C. Feeley, Sr., J. A. Otten, eds., American Society for Testing and Materials, Philadelphia, 1990.

[2] Brundage, J. F., R. Scott, W. M. Lednar, D. W. Smith and R. N. Miller, "Building-Associated Risk of Febrile Acute Respiratory Disease in Army Trainees," JAMA, 259, 2108-2112, 1988.

[2] Burge, S., A. Hedge, S. Wilson, J. H. Bass and A. Robertson, "Sick Building Syndrome: A Study of 4373 Office Workers," Ann. Occup. Hyg., 31, 493-504, 1987.

[4] Fanger, P.O., J. Lauridsen, P. Bluyssen and G. Clausen, "Air Pollution Sources in Offices and Assembly Halls, Quantified by the Olf Unit," Energy and Buildings, 12, 7-19, 1988.

[5] Morey, P.R., "Bioaerosols in the Indoor Environment: Current Practices and Approaches," Indoor Air Quality International Symposium," American Industrial Hygiene Association (in press).

[6] Feeley, J.C., Sr., "Legionellosis: Risk Associated With Building Design," In Architectural Design and Indoor Microbial Pollution, Ed. R. B. Kundsin, Oxford Univ. Press, New York, pp. 218-227, 1988.

[7] Morey, P. R. and J. C. Feeley, Sr., "Microbiological Aerosols Indoors," ASTM Standardization News, 16 (12), 54-58, 1988.

[8] Edwards, J. H., "Microbial and Immunological Investigations and Remedial Action After an Outbreak of Humidifier Fever," Bri. J. Ind. Med., 37, 55-62, 1980.

[9] Reed, C. E., M. C. Swanson, M. Lopez, A. M. Ford, J. Major, W. B. Witmer and T. B. Valdes, "Measurement of IgG Antibody and Airborne Antigen to Control an Industrial Outbreak of Hypersensitivity Pneumonitis," Jour. Occ. Med., 25, 207-210, 1983.

[10] American Society of Heating, Refrigerating and Air-conditioning Engineers Standard 62-1989, "Ventilation For Acceptable Indoor Air Quality", Atlanta, Georgia, 1989.

[11] Burge, H. A., J. C. Feeley, Sr., K. Kreiss, D. Milton, P. R. Morey, J. A. Otten, K. Peterson, J. J. Tulis, R. Tyndall, Guidelines for the Assessment of Bioaerosols in the Indoor Environment, American Conference of Governmental Industrial Hygienists, Cincinnati, 1989.

Hal Levin

ASTM THOUGHTS OF THE FUTURE AND CLOSING REMARKS

REFERENCE: Levin, H., "ASTM Thoughts of the Future and Closing Remarks," Biological Contmaninants in Indoor Environments. ASTM STP 1071, Philip R. Morey, James C. Feeley, Sr., James A. Otten, Editors, American Society for Testing and Materials, Philadelphia, 1990.

ABSTRACT: Changes in building design, construction, and operation during recent years have increased biological contamination of indoor air. The health risks versus the economic costs of controlling microorganisms in the indoor environment are discussed. The indoor air quality impacts of energy conservation and new features of construction are reviewed with reference to biological contaminants. Some potentially effective control measures and further research needs are identified. Potential litigation consideration are presented. Microbial contamination is discussed in the overall context of indoor environmental quality. Finally, the concept of "building ecology" is presented as a comprehensive approach to understanding the inter-relationships between building environments, building occupants, and the larger environment.

KEYWORDS: biological contaminants, microorganisms, indoor air quality, construction, building ecology

INTRODUCTION

Microbial Contaminants in the Overal Context of the Environment

The comments of microbiologists are strikingly anthropocentric to a non-microbiologist audience. Each symposium speaker described

Hal Levin is research architect with Hal Levin & Associates, 2548 Empire Grade, Santa Cruz, CA 95060.

microorganisms which they considered "good" or "bad" in relation to the known effects on humans (and, at times, on the accouterments of human civilization. Scientists concerned with controlling the harmful effects of a species on human life and health view its eradication as a positive accomplishment. This is implicitly a value-biased assessment; really a human bias.

In the past two or three decades, we (humans) have begun to invest large amounts of human and economic resources and forgo exploitation of certain natural resources in order selectively to preserve some threatened species. One of the rationales for such species preservation activities is the importance of ecological diversity. But we usually do not extend that argument to include species which seriously threaten human health. In fact, the use of pesticides demonstrates our willingness selectively to eliminate or eradicate species which cause us discomfort or indirect harm by damaging our crops or our buildings. But were would we be without the beneficial microorganisms in the human digestive tract or in the forest floor soil?

Narrowness and self-interest often drive these decisions. Why save the spotted owl or the African elephant or the Bald Eagle? Why not save certain bacteria or fungi? Why not preserve all species? These rhetorical questions are meant to illustrate our selectivity and the value-based biases of our social actions. What is a truly ecological perspective? Does it have any meaning without the anchor of a particular species perspective? After all, do any other species have programs for species conservation and protection? Has any other species ever been responsible for the eradication of so many others?

HEALTH RISK VERSUS ECONOMICS OF MICROBIAL CONTROL MEASURES

The symposium organizers assigned the present author the task of discussing the health risk versus economics of microbial control measures. The question requires knowledge and understanding of many subjects and the ability to apply them to the present question. First, we must understand the health effects of exposure to each of the mulititude of microorganims found indoors. Secondly, we must know what exposure will result in which health effect. Next, we must know how effectively to control each organism to eliminate the health risk. Finally, we must know the cost of implementing each control measure. Such knowledge is certainly beyond the qualifications of the present author.

The subject of health effects of the various microorganisms found indoor is, by itself, poorly understood. Many symposium speakers have said very little if anything about the health effects of exposure to the microorganims they discussed. Many of the organisms have not been previously regarded as hazards in indoor air, although there is evidence of exposure through airborne routes indoors. Many are believed to cause low-level symptoms similar to the symptoms associated with building-associated illnesses: sick building syndrome and building-related illness. Yet, the etiology of most building-associated illnesses is not well understood [1-2].

What have we learned about the health risks of microorganisms in indoor environments? Joseph Plouffe has told us several significant things.

1. We learn by disaster or tragedy - it takes an epidemic to get our attention focused on a problem or an organism.

2. We are proliferating conditions for microbial contamination, exposure, and infection (or allergy) at the same time as we are getting more clever (not necessarily smarter) with new instruments, techniques and more trained microbiologists. Significant changes in the environment have occurred which change the population exposure to microorganism-related illness. These include the following:

a. Increased efforts to conserve energy in buildings;
b. Population residence shift from private residences and hospitals to nursing homes;
c. Proliferation of air-conditioning in homes, cars, and increased use as population shifts to the "sun-belt;" and,
d. Increased environmental pollution attributable to population growth and industrial progress.

3. As time passes, the percentage of respiratory infections for which the etiology is unresolved has increased. In other words, either the world is changing; or as we learn more about it, we learn more about the limits of our understanding. The corrolary of the second possibility is that we learn how wrong we were in the past.

Having stated that neither the present author nor any one else knows enough about the health risks of most microorganisms in the indoor environment, let us consider the economics of controlling microbial contaminants. There are two elements to the questions: control measures and their economics.

Effective control measures are understood for only some of the many microorganisms. Nothing has been published that we are aware of that deals with the subject broadly. What little has been written looks at specific control alternatives for specific organisms or building components. That may be all we can (or should) do. However, control measures targeted at specific microorganims might have dynamic effects on non-target organisms resulting in secondary benefits or problems that should also be addressed.

The economics of broadly controlling microorganims in indoor environments is also a complex question. The introduction to this paper is more than a self-indulgent digression; it suggests that important value issues underly the bases for choices about the environment, indoors or outdoors.

The word "economics" shares its common root and some of its meaning with "ecology;" both deal with the resource flows in the environments of living organisms. Ecology comes from the Greek *oikos* (house) and *logy*, meaning science, knowledge or theory [3]. Economy comes from the Greek *oikonomika*, the science of household management [4]. Economics is the management of resources of an individual, a community or a country; it focuses on human-value based resources. Ecology is the broader term since it attempts to consider all resources of significance to all organisms in a particular environment (ecosystem).

Defining economics for evaluating practical alternatives requires explicit assumptions about the costs of failure to control the target organisms or diseases. In other words, predicting the likelihood and nature of disease and its consequences. We simply do not know enough to make such predictions.

Economics in its broadest sense is "home-making." For humans, the "home" is the environment *as we know and understand it*. In ecological terms, the challenge is to make a beneficial niche -- a healthy home, workplace, school, or other building.

In purely financial terms, economics is the discernible monetary cost of doing or not doing something; in the present case, the activity is controlling or not controlling indoor environmental microogranisms. All we can say is that the health risks will be reduced as microbial control measures increase. If we expend sufficient resources to microbial control, we might eliminate the health risks entirely. If we spend none, they might be large. This can be graphed conceptually as shown in Figure 1.

FIGURE 1 - Health Risks vs. Economics of Microbial Control Measures

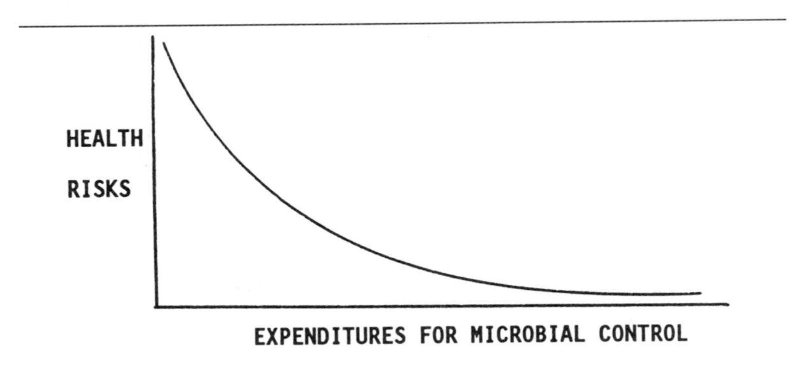

Our efforts to control microbial contaminants indoors will be enhanced by further research. But research priorities tend to be set in part as a response to public perceptions of risk. Relatively large commitments of resources to earthquake hazard mitigation followed the 1964 Anchorage earthquake and the 1971 San Fernando earthquake. As a result, significant gains in our understanding have been reflected in structural engineering, building codes, and earthquake disaster preparedness [5]. The serious outbreak of Legionnaire's Disease in Philadelphia in 1976 led to the scientific endeavors enabling identification of the organism and subsequent research on its control [6-7].

Public investments in research are closely related to public perception of risks and not necessarily related to the demonstrated magnitude of the hazard. Earthquakes tend to make the news when there is no human injury or death, and even when there is no significant property damage. The newsworthiness drives the perception of risk which, in turn, drives the commitment of research funds [8].

After a disaster, as Dr. Plouffe pointed out, there may be enough interest to produce some research funding. The research is likely to demonstrate that the risk is real but, perhaps, not as large as originally believed. As time passes after the disaster, the public will be less informed and less interested, and press coverage will diminish. The public will develop diverse views of the magnitude of risk, most likely diminishing over time. Scientists' views will become increasingly informed, and the perception of the risk will more closely reflect the measurable risk. Further funding is likely to depend upon recurrent episodes or new scientific findings of greater risk than previously believed.

In fact we use many, sometimes expensive control measures, for "economic" reasons when we perceive that not controlling indoor microorganisms will cost more than effective control measures. In industrial environments, measures are taken to control against microorganisms which might adversely affect the industrial activity. In commercial office buildings, biocides are routinely added to cooling tower water to prevent algae growth and to control Legionella. Fungicides are routinely added to paints, adhesives, carpets, composite wood products, and other building materials and furnishings. In health care facilities, managers watch for certain hazards which present known risks and can be controlled.

ENERGY CONSERVATION AND CONSTRUCTION

Energy Conservation Versus Build-up of Indoor Moisture

A major building design, operation and use issue is the conservation of energy. Ever since the Arab oil embargo of 1973, the building industry has focused on energy consumption. However, certain energy conservation measures, if not properly implemented, can result in humidity excursions outside the range recommended for control of microorganisms -- 30-60% RH [9].

One of the fundamental controls of humidity is regulation of air exchange with the outdoors. Indoor moisture levels depend on outside air ventilation rates and moisture content and indoor moisture sources and sinks. Climate will determine the nature of the effect, since outdoor air humidity can be above, below or near indoor humidity. Heating or cooling the ventilation air will change its moisture content and effect on indoor relative humidity. Figure 2 shows the relationships between indoor and outdoor humidity as a function of ventilation in buildings with typical human occupancy.

FIGURE 2 - Indoor/Outdoor Relative Humidity Ratio as a Function of Ventilation in Various Climates

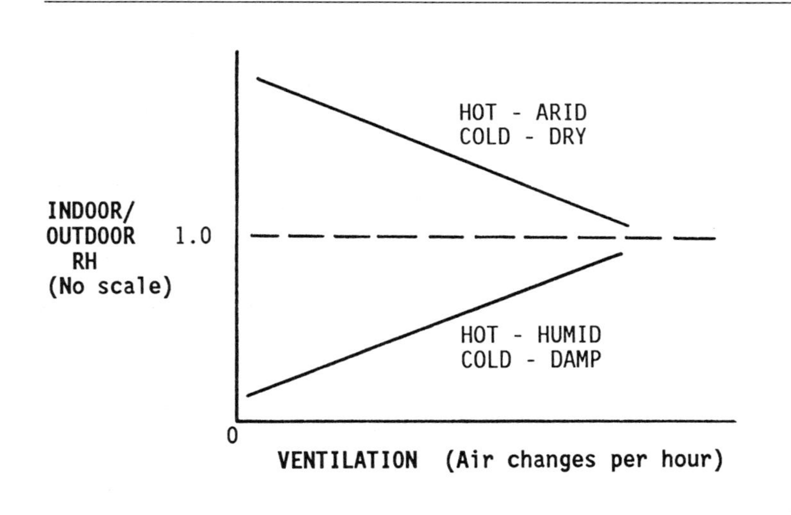

Certain types of heat exchangers used to conserve energy while providing building air exchange with the outdoors can maintain humidity. However, precautions must be taken to prevent deterioration of the heat exchange device or adverse air quality effects from retention of water soluble contaminants such as formaldehyde.

NEW FEATURES IN CONSTRUCTION

Alternatives to Man-Made Fibrous Insulation in HVAC Systems

Many researchers have identified man-made fibrous insulations used in building HVAC systems as important substrates for microorganism proliferation [10]. Either by providing extended surface areas or nutrients, or both, fibrous insulation materials often become very contaminated. Recent research has shown that when these materials become dirty, they much more readily become moist [11]. When both moist and dirty, they can provide hospitable environments for many common bioaerosols. However, without some sort of sound attenuation, HVAC systems can produce noise which reduces the suitability of the occupied spaces. And thermal insulation is necessary for effective and economical operation of the systems. Therefore, it is important that alternative materials or means of providing thermal and acoustic control in HVAC systems be developed and used.

One common approach is to place fibrous insulation materials on the outside rather than inside the ductwork. Thus, the air flow is not directly exposed to the fibrous materials. However, where the supply air ducts are located above a suspended ceiling used to enclose a return air plenum, then return air does flow over the duct insulation. In that case, and especially where insulation is placed inside the duct, a vapor barrier should be used to separate the fibrous material from the circulating air. Vapor barriers covering acoustic insulation reduce the effectiveness of the insulation, and this must be considered in the design.

Another approach is to coat the fibrous insulations with materials which protect the surface of the fibers from erosion by the air movement. These materials might be useful in isolating the fibers from the air stream if their application was done for that purpose. Microscopic analysis of some coated materials reveals that the fibrous materials are often not adequately separated by the coating, although it may be applied heavily enough to reduce surface erosion [12].

Fibrous liners used inside air mixing chambers or near fans are often used for acoustic control. Ventilation fans make a great deal of noise which can be transmitted through the ductwork to the occupied spaces. Therefore, liners are often used upstream and downstream of return air and supply air fans respectively to absorb some of the emitted sound. Alternatively or in addition, sound traps are used. The sound traps themselves may be constructed of fibrous materials.

Replacement of Porous Microbial Reservoirs in Buildings with Acceptable Alternatives

Thus far there have not been new materials introduced to replace many of the fibrous materials which enjoy widespread use for a multitude of building applications. One of the major uses is for sound control. Carpeting has become standard in homes, offices, schools, restaurants, and other public spaces. A major reason for its use is to absorb sound. Yet carpets are often discovered to be the source of microbial contamination indoors. Suspended ceilings are also extensively used for sound control. However, there are other approaches to acoustic control which involve architectural design solutions rather than surface materials.

One visually exciting example is the ceiling over many portions of the public spaces in the Toronto City Hall. The ceiling consists of regularly spaced metallic slats in the ceiling plane with open spaces between each pair of slats. This allows some of the sound waves to pass above the ceiling where they can be reflected and absorbed. Other design approaches have been used for many years in restaurants, auditoria, sports facilities, and other environments where noise control is important. Growing interest in reducing potential reservoirs and amplifiers for microbial contaminants indoors will increase future use of new approaches.

Direct Digital Control (DDC)

Direct digital control of HVAC systems will allow input from multiple sensors to a computer or to a microprocessor located in a local ventilation zone. Sensors for relative humidity will become more common in the very near future, and the input will assist not only in avoiding excess humidity but also in better achieving energy conservation and comfort [13].

Increased occupant control: As indoor environmental control has become tighter to achieve energy conservation, growing numbers of occupants are becoming dissatisfied with the conditions in their work environment. Some designers and building equipment manufacturers are currently interested in providing more direct control of their environment to building occupants. This has become more practical with recently developed technologies to accomplish it. Since different individual's responses to a given set of environmental conditions will vary significantly, it is more attractive to designers and building operators to provide more individual occupant control [14]. Some set of parameters is programmed into a microprocessor control module which can adjust environmental conditions according to occupant instructions. The system works somewhat like automobile air-conditioning.

One major furniture manufacturer has introduced a work station system for the open office environment that gives occupants varying degrees of control over lighting, ventilation, and acoustic conditions. Other companies are working on similar systems. Again, the direct digital control (DDC) systems will allow integration of multiple inputs and production of a more comfortable and habitable building environment.

Biocides

New biocides are being marketed which are integrated into the products, especially carpeting. The manufacturers claim these biocides are "permanently" attached to the fibers they protect, therefore eliminating concerns about exposure of building occupants to toxic chemicals. If these biocides are effective and provide long-term protection, they can significantly reduce the amplification of bioaerosols indoors.

Air Cleaning

Air cleaning equipment in buildings will evolve during the next few years to meet the requirements of the new generation ASHRAE ventilation standard (Standard 62-1989, Ventilation for Acceptable Indoor Air Quality). More and more local and state jurisdictions are adopting requirements for implementation of the ASHRAE standard, not only as a design standard but also as an operational standard for occupied buildings. The air cleaner industry is poised and waiting to exploit the expanded market for products that the standard will engender.

Filter problems: The move to more air cleaning to meet indoor air quality standards, guidelines and codes will result in more and different types of filtration. Filter media are often found to be contaminated by microorganisms and suspected of being bioaerosol sources. New methods for identifying potential contamination, preventing it, and controlling it will be required. Biocides, inspections, purging, and new materials are all potential control measures that may find varying degrees of use.

Humidification

Humidification of building air is a major source of concern for those interested in control of microorganisms [15]. Several types of humidification equipment are employed with varying degrees of economy, efficacy, and potential for bioaerosol generation.

Portable humidifiers: Household and personal humidifiers often contain water reservoirs which support microbial growth if not carefully and diligently maintained. Since most domestic water supplies contain some microorganisms, it is not surprising that tap water is often advised against by manufacturers, consumer agencies, and public health officials [16]. Daily cleaning and changing of water reservoirs are typical recommendations.

Humidication of commercial facilities: Humidifiers used in large HVAC systems can be bioaerosol sources if not properly controlled [17]. Design, maintenance, and operational considerations apply.

LITIGATION

What Actions Should a Building Operator or Industrial Operator Carry out in Order to Protect Onself from Litigation?

Litigation over indoor air quality complaints and illnesses is becoming increasingly common. Building owners and operators are looking for ways to reduce their liability. Consultants are being involved during the design phase to minimize risks and avoid obvious problems. Specialists are being hired to do routine investigations and to certify "healthy buildings." And duct cleaning, more efficient filters, and overall maintenance practices are improving where building operators are concerned about litigation.

Designers and manufacturers are being held accountable by their clients for indoor air quality. This means they are spending more effort on identifying and minimizing risks of microbial contamination.

Practical limitations: There are a multitude of organisms and situations which warrant attention. The building development process is complex and the number and types of participants is large. It is not possible to anticipate all eventualiaties. However, careful consideration of microorganism control at each stage of a building's life can reduce singificantly the likelihood that air quality problems will occur.

MICROBIALS IN THE OVERALL CONTEXT OF INDOOR ENVIRONMENTAL QUALITY

Experts on a World Health Organization Working Group on indoor air quality rank ordered the hazards. A modified and updated ranking is shown in Table 1. Microbial contaminants have been moving up the list in recent years as more and more indoor air investigators find or hypothesize microorgansim contamination in sick building syndrome and building-related illness. Additional knowledge about the selective amplification of Legionella pneumophila and other infective microorganisms has resulted in increased awareness, concern, collection, and attribution of causality.

TABLE 1 - Indoor air quality contaminants of concern ranked in order of presumed importance to public health

Contaminant
Environmental tobacco smoke (ETS)
Radon
Asbestos
Organic chemicals (VOC)
Microbial contaminants
NO_2
Ozone
Particulates

Buildings as Systems

Many investigators have failed to diagnose problems in buildings because they have not approached the building as a dynamic, complex system. Research and investigatory techniques are becoming more sophisticated as our understanding of the nature of indoor air quality problems evolves. Many scientific and professional disciplines have become involved in indoor air quality investigations [18]. The diagnosis is often a function of the disciplinary bias of the investigators rather than the result of a comprehensive, multi-disciplinary or inter-disciplinary approach [19].

Building Ecology

We have coined the term "building ecology" to describe the study of building environments as systems with complex interrelationships to the occupants and to the larger environment [20]. We have borrowed the term "ecology" from the branch of biology which examines ecological systems by studying the populations, energy and nutritional flows, and habitats. It is our view that building environments must be treated with a similarly systematic, scientific approach before we will gain a firm grasp on the causes and remedies of building associated illnesses. We must learn from the ecologists and view buildings as both complex and dynamic systems before we can obtain the understanding we need to create healthy buildings.

Scientists must cooperate with architects, engineers, and building operators to develop a much more comprehensive approach to investigating problem buildings and creating healthy ones. Improved sampling and analytical methods for microbial contaminants in indoor air are an essential element of the needed science. ASTM's Subcommittee D22.05 on Indoor Air will continue to provide a forum for the development and standardization of these methods and the related knowledge.

REFERENCES

[1] Hodgson, M.J. and Kreiss, K., "Building-Associated Diseases: An Update." in Proceedings of IAQ '86: Managing Indoor Air for Health and Energy Conservation. American Society for Heating, Refrigerating and Air-Conditioning Engineers, Inc., Atlanta, 1986. pp. 1-15.
[2] Kreiss, K. and Hodgson, M.J., "Building-Associated Epidemics." in Walsh, P.J., Dudney, C.S., and Copenhaver, E.D., Indoor Air Quality. CRC Press, Inc., Boca Raton, Florida, 1984. pp. 87-108.
[3] Webster, N. Webster's New Universal Unabridged Dictionary, Second Edition. Simon & Schuster, New York, 1983.
[4] Webster, Ibid
[5] Schodek, D.L., "Research on Natural and Man-Made Hazards: Impacts on Building Regulations." in Cooke, P., ed., Research and Innovation in the Building Regulatory Process. NBS Special Publication 518. National Bureau of Standards, Washington, D.C., 1977, pp. 25-47.

[6] American Society for Microbiology, Legionella: Proceedings of the 2nd International Symposium. American Society for Microbiology, Washington, D. C., 1987.
[7] Fraser, D.W., and McDade, J. E., "Legionellosis." Scientific American, Vol. 241, 1979, pp. 82-99.
[8] Burton, I., Kales, R.W. and White, G.F., The Environment as Hazard. Oxford University Press, New York, 1978.
[9] Morey, P., "Microorganisms in Buildings and HVAC Systems: A Summary of 21 Environmental Studies." in Engineering Solutions to Indoor Air Problems; Proceedings of IAQ '87. American Society for Heating, Refrigerating and Air-Conditioning Engineers, Inc., Atlanta, 1987, pp. 10-24.
[10] Morey Ibid
[11] West, M.K., and E. C. Hansen, "Determination of Material hygroscopic Properties Which Affect Indoor Air Quality." The Human Equation: Health and Comfort, Porceedings of IAQ '89. American Society for Heating, Refrigerating and Air-Conditioning Engineers, Inc., Atlanta, 1990.
[12] West and Hansen, Ibid
[13] Kroner, W.M., ed. A New Frontier: Environments for Innovation; Proceedings of the International Symposium on Advanced Comfort Systems for the Work Environment. Troy, N.Y.: Center for Architectural Research, Rensselaer Polytechnic Institute, 1988. 394 pp.
[14] Rohles, F. 1988. "Comfort in the Man-Environment System." in Kroner, W.M., ed. Op. Cit., pp 295-302.
[15] Morey Ibid
[16] U.S. Consumer Product Safety Commission, "Consumer Alert," 1988.
[17] Morey Ibid
[18] Woods, J.E., Morey, P.R., and Rask, D.R., "Indoor Air Diagnostics: Qualitative and Quantitative Procedures to Improve Environmental Conditions." in Nagda, N.L., and Harper, J., Design and Protocol for Monitoring Indoor Air Quality, STP 1002. Philadelphia: American Society for Testing and Materials, 1989, pp. 80-98.
[19] Molhave, L., "The Sick Buildings -- A Subpopulation Among the Problem Buildings." in Seifert, B., ed., Indoor Air '87; Proceedings of the 4th International Conference on Indoor Air Quality and Climate, Volume 2, . Berlin: Institute for Water, Soil and Air Hygiene, 1987, pp. 469-473.
[20] Levin, H. "Building Ecology." Progressive Architecture. April 1981.

Author Index

Subject Index